Pitman Research Notes in Mathematics Series

Submission of proposals for consideration

Suggestions for publication, in the form of outlines and representative samples, are invited by the Editorial Board for assessment. Intending authors should approach one of the main editors or another member of the Editorial Board, citing the relevant AMS subject classifications. Alternatively, outlines may be sent directly to the publisher's offices. Refereeing is by members of the board and other mathematical authorities in the topic concerned, throughout the world.

Preparation of accepted manuscripts

On acceptance of a proposal, the publisher will supply full instructions for the preparation of manuscripts in a form suitable for direct photo-lithographic reproduction. Specially printed grid sheets can be provided and a contribution is offered by the publisher towards the cost of typing. Word processor output, subject to the publisher's approval, is also acceptable.

Illustrations should be prepared by the authors, ready for direct reproduction without further improvement. The use of hand-drawn symbols should be avoided wherever possible, in order to maintain maximum clarity of the text.

The publisher will be pleased to give any guidance necessary during the preparation of a typescript, and will be happy to answer any queries.

Important note

In order to avoid later retyping, intending authors are strongly urged not to begin final preparation of a typescript before receiving the publisher's guidelines. In this way it is hoped to preserve the uniform appearance of the series.

Longman Scientific & Technical
Longman House
Burnt Mill
Harlow, Essex, CM20 2JE
UK
(Telephone (0279) 426721)

Titles in this series. A full list is available from the publisher on request.

100 Optimal control of variational inequalities
V Barbu
101 Partial differential equations and dynamical systems
W E Fitzgibbon III
102 Approximation of Hilbert space operators Volume II
C Apostol, L A Fialkow, D A Herrero and D Voiculescu
103 Nondiscrete induction and iterative processes
V Ptak and F-A Potra
104 Analytic functions – growth aspects
O P Juneja and G P Kapoor
105 Theory of Tikhonov regularization for Fredholm equations of the first kind
C W Groetsch
106 Nonlinear partial differential equations and free boundaries. Volume I
J I Díaz
107 Tight and taut immersions of manifolds
T E Cecil and P J Ryan
108 A layering method for viscous, incompressible L_p flows occupying R^n
A Douglis and E B Fabes
109 Nonlinear partial differential equations and their applications: Collège de France Seminar. Volume VI
H Brezis and J L Lions
110 Finite generalized quadrangles
S E Payne and J A Thas
111 Advances in nonlinear waves. Volume II
L Debnath
112 Topics in several complex variables
E Ramírez de Arellano and D Sundararaman
113 Differential equations, flow invariance and applications
N H Pavel
114 Geometrical combinatorics
F C Holroyd and R J Wilson
115 Generators of strongly continuous semigroups
J A van Casteren
116 Growth of algebras and Gelfand–Kirillov dimension
G R Krause and T H Lenagan
117 Theory of bases and cones
P K Kamthan and M Gupta
118 Linear groups and permutations
A R Camina and E A Whelan
119 General Wiener–Hopf factorization methods
F-O Speck
120 Free boundary problems: applications and theory. Volume III
A Bossavit, A Damlamian and M Fremond
121 Free boundary problems: applications and theory. Volume IV
A Bossavit, A Damlamian and M Fremond
122 Nonlinear partial differential equations and their applications: Collège de France Seminar. Volume VII
H Brezis and J L Lions
123 Geometric methods in operator algebras
H Araki and E G Effros
124 Infinite dimensional analysis–stochastic processes
S Albeverio
125 Ennio de Giorgi Colloquium
P Krée
126 Almost-periodic functions in abstract spaces
S Zaidman
127 Nonlinear variational problems
A Marino, L Modica, S Spagnolo and M Degliovanni
128 Second-order systems of partial differential equations in the plane
L K Hua, W Lin and C-Q Wu
129 Asymptotics of high-order ordinary differential equations
R B Paris and A D Wood
130 Stochastic differential equations
R Wu
131 Differential geometry
L A Cordero
132 Nonlinear differential equations
J K Hale and P Martinez-Amores
133 Approximation theory and applications
S P Singh
134 Near-rings and their links with groups
J D P Meldrum
135 Estimating eigenvalues with *a posteriori/a priori* inequalities
J R Kuttler and V G Sigillito
136 Regular semigroups as extensions
F J Pastijn and M Petrich
137 Representations of rank one Lie groups
D H Collingwood
138 Fractional calculus
G F Roach and A C McBride
139 Hamilton's principle in continuum mechanics
A Bedford
140 Numerical analysis
D F Griffiths and G A Watson
141 Semigroups, theory and applications. Volume I
H Brezis, M G Crandall and F Kappel
142 Distribution theorems of L-functions
D Joyner
143 Recent developments in structured continua
D De Kee and P Kaloni
144 Functional analysis and two-point differential operators
J Locker
145 Numerical methods for partial differential equations
S I Hariharan and T H Moulden
146 Completely bounded maps and dilations
V I Paulsen
147 Harmonic analysis on the Heisenberg nilpotent Lie group
W Schempp
148 Contributions to modern calculus of variations
L Cesari
149 Nonlinear parabolic equations: qualitative properties of solutions
L Boccardo and A Tesei
150 From local times to global geometry, control and physics
K D Elworthy

151 A stochastic maximum principle for optimal control of diffusions
U G Haussmann
152 Semigroups, theory and applications. Volume II
H Brezis, M G Crandall and F Kappel
153 A general theory of integration in function spaces
P Muldowney
154 Oakland Conference on partial differential equations and applied mathematics
L R Bragg and J W Dettman
155 Contributions to nonlinear partial differential equations. Volume II
J I Díaz and P L Lions
156 Semigroups of linear operators: an introduction
A C McBride
157 Ordinary and partial differential equations
B D Sleeman and R J Jarvis
158 Hyperbolic equations
F Colombini and M K V Murthy
159 Linear topologies on a ring: an overview
J S Golan
160 Dynamical systems and bifurcation theory
M I Camacho, M J Pacifico and F Takens
161 Branched coverings and algebraic functions
M Namba
162 Perturbation bounds for matrix eigenvalues
R Bhatia
163 Defect minimization in operator equations: theory and applications
R Reemtsen
164 Multidimensional Brownian excursions and potential theory
K Burdzy
165 Viscosity solutions and optimal control
R J Elliott
166 Nonlinear partial differential equations and their applications: Collège de France Seminar. Volume VIII
H Brezis and J L Lions
167 Theory and applications of inverse problems
H Haario
168 Energy stability and convection
G P Galdi and B Straughan
169 Additive groups of rings. Volume II
S Feigelstock
170 Numerical analysis 1987
D F Griffiths and G A Watson
171 Surveys of some recent results in operator theory. Volume I
J B Conway and B B Morrel
172 Amenable Banach algebras
J-P Pier
173 Pseudo-orbits of contact forms
A Bahri
174 Poisson algebras and Poisson manifolds
K H Bhaskara and K Viswanath
175 Maximum principles and eigenvalue problems in partial differential equations
P W Schaefer
176 Mathematical analysis of nonlinear, dynamic processes
K U Grusa
177 Cordes' two-parameter spectral representation theory
D F McGhee and R H Picard
178 Equivariant K-theory for proper actions
N C Phillips
179 Elliptic operators, topology and asymptotic methods
J Roe
180 Nonlinear evolution equations
J K Engelbrecht, V E Fridman and E N Pelinovski
181 Nonlinear partial differential equations and their applications: Collège de France Seminar. Volume IX
H Brezis and J L Lions
182 Critical points at infinity in some variational problems
A Bahri
183 Recent developments in hyperbolic equations
L Cattabriga, F Colombini, M K V Murthy and S Spagnolo
184 Optimization and identification of systems governed by evolution equations on Banach space
N U Ahmed
185 Free boundary problems: theory and applications. Volume I
K H Hoffmann and J Sprekels
186 Free boundary problems: theory and applications. Volume II
K H Hoffmann and J Sprekels
187 An introduction to intersection homology theory
F Kirwan
188 Derivatives, nuclei and dimensions on the frame of torsion theories
J S Golan and H Simmons
189 Theory of reproducing kernels and its applications
S Saitoh
190 Volterra integrodifferential equations in Banach spaces and applications
G Da Prato and M Iannelli
191 Nest algebras
K R Davidson
192 Surveys of some recent results in operator theory. Volume II
J B Conway and B B Morrel
193 Nonlinear variational problems. Volume II
A Marino and M K V Murthy
194 Stochastic processes with multidimensional parameter
M E Dozzi
195 Prestressed bodies
D Iesan
196 Hilbert space approach to some classical transforms
R H Picard
197 Stochastic calculus in application
J R Norris
198 Radical theory
B J Gardner
199 The C^*-algebras of a class of solvable Lie groups
X Wang
200 Stochastic analysis, path integration and dynamics
K D Elworthy and J C Zambrini

201 Riemannian geometry and holonomy groups
S Salamon
202 Strong asymptotics for extremal errors and polynomials associated with Erdös type weights
D S Lubinsky
203 Optimal control of diffusion processes
V S Borkar
204 Rings, modules and radicals
B J Gardner
205 Two-parameter eigenvalue problems in ordinary differential equations
M Faierman
206 Distributions and analytic functions
R D Carmichael and D Mitrovic
207 Semicontinuity, relaxation and integral representation in the calculus of variations
G Buttazzo
208 Recent advances in nonlinear elliptic and parabolic problems
P Bénilan, M Chipot, L Evans and M Pierre
209 Model completions, ring representations and the topology of the Pierce sheaf
A Carson
210 Retarded dynamical systems
G Stepan
211 Function spaces, differential operators and nonlinear analysis
L Paivarinta
212 Analytic function theory of one complex variable
C C Yang, Y Komatu and K Niino
213 Elements of stability of visco-elastic fluids
J Dunwoody
214 Jordan decomposition of generalized vector measures
K D Schmidt
215 A mathematical analysis of bending of plates with transverse shear deformation
C Constanda
216 Ordinary and partial differential equations. Volume II
B D Sleeman and R J Jarvis
217 Hilbert modules over function algebras
R G Douglas and V I Paulsen
218 Graph colourings
R Wilson and R Nelson
219 Hardy-type inequalities
A Kufner and B Opic
220 Nonlinear partial differential equations and their applications: Collège de France Seminar. Volume X
H Brezis and J L Lions
221 Workshop on dynamical systems
E Shiels and Z Coelho
222 Geometry and analysis in nonlinear dynamics
H W Broer and F Takens
223 Fluid dynamical aspects of combustion theory
M Onofri and A Tesei
224 Approximation of Hilbert space operators. Volume I. 2nd edition
D Herrero
225 Operator theory: proceedings of the 1988 GPOTS–Wabash conference
J B Conway and B B Morrel
226 Local cohomology and localization
J L Bueso Montero, B Torrecillas Jover and A Verschoren
227 Nonlinear waves and dissipative effects
D Fusco and A Jeffrey
228 Numerical analysis 1989
D F Griffiths and G A Watson
229 Recent developments in structured continua. Volume II
D De Kee and P Kaloni
230 Boolean methods in interpolation and approximation
F J Delvos and W Schempp
231 Further advances in twistor theory. Volume I
L J Mason and L P Hughston
232 Further advances in twistor theory. Volume II
L J Mason and L P Hughston
233 Geometry in the neighborhood of invariant manifolds of maps and flows and linearization
U Kirchgraber and K Palmer
234 Quantales and their applications
K I Rosenthal
235 Integral equations and inverse problems
V Petkov and R Lazarov
236 Pseudo-differential operators
S R Simanca
237 A functional analytic approach to statistical experiments
I M Bomze
238 Quantum mechanics, algebras and distributions
D Dubin and M Hennings
239 Hamilton flows and evolution semigroups
J Gzyl
240 Topics in controlled Markov chains
V S Borkar
241 Invariant manifold theory for hydrodynamic transition
S Sritharan
242 Lectures on the spectrum of $L^2(\Gamma\backslash G)$
F L Williams
243 Progress in variational methods in Hamiltonian systems and elliptic equations
M Girardi, M Matzeu and F Pacella
244 Optimization and nonlinear analysis
A Ioffe, M Marcus and S Reich
245 Inverse problems and imaging
G F Roach
246 Semigroup theory with applications to systems and control
N U Ahmed
247 Periodic-parabolic boundary value problems and positivity
P Hess
248 Distributions and pseudo-differential operators
S Zaidman
249 Progress in partial differential equations: the Metz surveys
M Chipot and J Saint Jean Paulin
250 Differential equations and control theory
V Barbu

251 Stability of stochastic differential equations with respect to semimartingales
X Mao
252 Fixed point theory and applications
J Baillon and M Théra
253 Nonlinear hyperbolic equations and field theory
M K V Murthy and S Spagnolo
254 Ordinary and partial differential equations. Volume III
B D Sleeman and R J Jarvis
255 Harmonic maps into homogeneous spaces
M Black
256 Boundary value and initial value problems in complex analysis: studies in complex analysis and its applications to PDEs 1
R Kühnau and W Tutschke
257 Geometric function theory and applications of complex analysis in mechanics: studies in complex analysis and its applications to PDEs 2
R Kühnau and W Tutschke
258 The development of statistics: recent contributions from China
X R Chen, K T Fang and C C Yang
259 Multiplication of distributions and applications to partial differential equations
M Oberguggenberger
260 Numerical analysis 1991
D F Griffiths and G A Watson
261 Schur's algorithm and several applications
M Bakonyi and T Constantinescu
262 Partial differential equations with complex analysis
H Begehr and A Jeffrey
263 Partial differential equations with real analysis
H Begehr and A Jeffrey
264 Solvability and bifurcations of nonlinear equations
P Drábek
265 Orientational averaging in mechanics of solids
A Lagzdins, V Tamuzs, G Teters and A Kregers
266 Progress in partial differential equations: elliptic and parabolic problems
C Bandle, J Bemelmans, M Chipot, M Grüter and J Saint Jean Paulin
267 Progress in partial differential equations: calculus of variations, applications
C Bandle, J Bemelmans, M Chipot, M Grüter and J Saint Jean Paulin
268 Stochastic partial differential equations and applications
G Da Prato and L Tubaro
269 Partial differential equations and related subjects
M Miranda
270 Operator algebras and topology
W B Arveson, A S Mishchenko, M Putinar, M A Rieffel and S Stratila
271 Operator algebras and operator theory
W B Arveson, A S Mishchenko, M Putinar, M A Rieffel and S Stratila
272 Ordinary and delay differential equations
J Wiener and J K Hale
273 Partial differential equations
J Wiener and J K Hale
274 Mathematical topics in fluid mechanics
J F Rodrigues and A Sequeira
275 Green functions for second order parabolic integro-differential problems
M G Garroni and J F Menaldi
276 Riemann waves and their applications
M W Kalinowski
277 Banach C(K)-modules and operators preserving disjointness
Y A Abramovich, E L Arenson and A K Kitover
278 Limit algebras: an introduction to subalgebras of C*-algebras
S C Power
279 Abstract evolution equations, periodic problems and applications
D Daners and P Koch Medina
280 Emerging applications in free boundary problems
J Chadam and H Rasmussen
281 Free boundary problems involving solids
J Chadam and H Rasmussen
282 Free boundary problems in fluid flow with applications
J Chadam and H Rasmussen
283 Asymptotic problems in probability theory: stochastic models and diffusions on fractals
K D Elworthy and N Ikeda
284 Asymptotic problems in probability theory: Wiener functionals and asymptotics
K D Elworthy and N Ikeda
285 Dynamical systems
R Bamon, R Labarca, J Lewowicz and J Palis
286 Models of hysteresis
A Visintin
287 Moments in probability and approximation theory
G A Anastassiou
288 Mathematical aspects of penetrative convection
B Straughan
289 Ordinary and partial differential equations. Volume IV
B D Sleeman and R J Jarvis
290 *K*-theory for real *C**-algebras
H Schröder
291 Recent developments in theoretical fluid mechanics
G P Galdi and J Necas
292 Propagation of a curved shock and nonlinear ray theory
P Prasad
293 Non-classical elastic solids
M Ciarletta and D Ieşan
294 Multigrid methods
J Bramble
295 Entropy and partial differential equations
W A Day
296 Progress in partial differential equations: the Metz surveys 2
M Chipot
297 Nonstandard methods in the calculus of variations
C Tuckey
298 Barrelledness, Baire-like- and (LF)-spaces
M Kunzinger
299 Nonlinear partial differential equations and their applications. Collège de France Seminar. Volume XI
H Brezis and J L Lions
300 Introduction to operator theory
T Yoshino

301 Generalized fractional calculus and applications
V Kiryakova
302 Nonlinear partial differential equations and their applications. Collège de France Seminar Volume XII
303 Numerical analysis 1993
D F Griffiths and G A Watson
304 Topics in abstract differential equations
S Zaidman
305 Complex analysis and its applications
C C Yang, G C Wen, K Y Li and Y M Chiang
306 Computational methods for fluid-structure interaction
J M Crolet and R Ohayon
307 Random geometrically graph directed self-similar multifractals
L Olsen
308 Progress in theoretical and computational fluid mechanics
G P Galdi, J Málek and J Necas
309 Variational methods in Lorentzian geometry
A Masiello
310 Stochastic analysis on infinite dimensional spaces
H Kunita and H-H Kuo
311 Representations of Lie groups and quantum groups
V Baldoni and M Picardello
312 Common zeros of polynomials in several variables and higher dimensional quadrature
Y Xu

Yuan Xu
University of Oregon, USA

Common zeros of polynomials in several variables and higher dimensional quadrature

CRC Press
Taylor & Francis Group
Boca Raton London New York

CRC Press is an imprint of the
Taylor & Francis Group, an **informa** business
A CHAPMAN & HALL BOOK

CRC Press
Taylor & Francis Group
6000 Broken Sound Parkway NW, Suite 300
Boca Raton, FL 33487-2742

First issued in hardback 2018

ISBN 13: 978-1-138-41773-1 (hbk)
ISBN 13: 978-0-582-24670-6 (pbk)

Visit the Taylor & Francis Web site at
http://www.taylorandfrancis.com

and the CRC Press Web site at
http://www.crcpress.com

To Litian

TABLE OF CONTENTS

Preface

A polynomial of degree n in one variable has n (possibly complex) zeros. For an orthogonal polynomial more is known: all zeros are real and distinct. One important application of this fact is in numerical integration. For a nonnegative measure on $\mathbb{R}$ with finite moments and infinite support, a Gaussian quadrature formula exists if, and only if, its nodes are the zeros of a corresponding orthogonal polynomial.

For a polynomial of two variables, a zero can mean either a point or a curve on the plane. In general, zeros of polynomials in d variables are algebraic varieties of dimension $d-1$ or less; they are usually difficult to deal with. On the other hand, there are more than one linearly independent polynomials of a fixed degree in several variables; for example, in two variables there are $n+1$ independent polynomials of degree n. From the point of view of quadrature, it is natural to study, instead of zeros of a single polynomial, common zeros of all polynomials of degree n, which could be distinct and real.

The purpose of this monograph is to study the common zeros of families of polynomials in several variables which are related to higher dimensional quadrature. We will introduce a new approach and use it to conduct a systematic study. The monograph is in essence a research paper; a good portion of the theorems are new and, in many cases, new proofs are given for known results. Nevertheless, the development is basically self-contained and the monograph can be taken as an introduction to this topic.

This work was supported by the National Science Foundation under Grant No. 9302721 and the Alexander von Humboldt-Foundation. Most of the work was carried out while I was visiting the University of Erlangen-Nuremberg, supported by an Alexander von Humboldt Fellowship. It is my pleasure to thank Professor Hubert Berens for his hospitality and many stimulating discussions; I also thank him for the great care with which he went through the entire monograph and suggested numerous changes that greatly improve the presentation.

Yuan Xu

1. Introduction

1.1. Review of the theory in one variable

The relation between Gaussian quadrature and zeros of orthogonal polynomials are well-known. To motivate our discussion in the setting of several variables, we shall briefly recall the basics of one variable theory (cf. [4, 32]).

Let μ be a nonnegative measure with finite moments and infinite support. Let $\mathbb{N}$ be the set of positive integers and let $n, m \in \mathbb{N}$. With respect to μ a linear functional

$$I_n(f) = \sum_{k=1}^{n} \lambda_k f(x_k), \qquad \lambda_k \in \mathbb{R}, \quad x_k \in \mathbb{R},$$

is called an (m, n) quadrature formula, which has n nodes and degree m, if

$$I_n(f) = \int_{\mathbb{R}} f(x)\, d\mu, \qquad \forall f \in \Pi_m,$$

and if there exists at least one element of Π_{m+1} such that equality fails to hold. Here and in the following, Π_n denotes the space of polynomials of degree at most n in one variable. The x_k's are called nodes and the λ_k's weights of the quadrature. An (m, n) quadrature is called positive if all λ_k are positive. For a fixed n it is well-known that $m \le 2n - 1$ for every (m, n) quadrature formula; the $(2n - 1, n)$ quadrature, which has the highest precision, is the so-called Gaussian quadrature which is positive.

Let $\{p_n\}_{n=0}^{\infty}$ be the system of orthonormal polynomials with respect to μ; *i.e.*,

$$\int_{-\infty}^{\infty} p_n(x) p_m(x)\, d\mu = \delta_{m,n} \quad \text{and} \quad p_n(x) = \gamma_n x^n + \ldots \in \Pi_n.$$

It is well-known that there exist numbers $a_n, b_n \in \mathbb{R}$ such that $\{p_n\}_{n=0}^{\infty}$ satisfies a three-term relation

(1.1.1) $$x p_n = a_n p_{n+1} + b_n p_n + a_{n-1} p_{n-1}, \quad n \ge 0, \quad a_n = \gamma_n / \gamma_{n+1},$$

where p_{-1} is defined to be zero; in fact,

$$a_n = \int_{-\infty}^{\infty} x\, p_n(x)\, p_{n+1}(x)\, dx \qquad \text{and} \qquad b_n = \int_{-\infty}^{\infty} x\, p_n^2(x)\, dx.$$

Moreover, Favard's theorem states that a sequence of polynomials is orthogonal with respect to a nonnegative Borel measure if, and only if, it satisfies a three-term relation with $a_n > 0$ for all n. The coefficients of (1.1.1) define the associated Jacobi matrix which is an infinite tridiagonal matrix, whose $n \times n$ submatrix at the left upper corner is given as

$$J_n = \begin{bmatrix} b_0 & a_0 & & & \bigcirc \\ a_0 & b_1 & a_1 & & \\ & \ddots & \ddots & \ddots & \\ & & a_{n-3} & b_{n-2} & a_{n-2} \\ \bigcirc & & & a_{n-2} & b_{n-1} \end{bmatrix}.$$

Using the matrix, p_n can be explicitly given as

$$p_n = \gamma_n x^n + \ldots = \gamma_n \det(xI_n - J_n), \tag{1.1.2}$$

where I_n is the identity matrix. In particular, this shows that zeros of p_n are eigenvalues of J_n which are clearly real. Moreover, it can be shown that all zeros of p_n are distinct.

The relation between Gaussian quadrature and zeros of orthogonal polynomials is as follows: an (m, n) quadrature formula is Gaussian, that is, $m = 2n - 1$, if and only if its nodes are the zeros of p_n. Moreover, there is a relation between the (m, n), $m > n$, positive quadrature and *quasi-orthogonal* polynomials. For example, it is well-known (cf. [32]) that the zeros of a quasi-orthogonal polynomial $q_{n,1} = p_n + \rho_1 p_{n-1}$ are the nodes of a $(2n-2, n)$ quadrature formula. More generally, let $r \in \mathrm{N}$, $r \leq n$. A quasi-orthogonal polynomial of degree n and order r is defined by

$$q_{n,r} = p_n + \rho_1 p_{n-1} + \ldots \rho_r p_{n-r}, \tag{1.1.3}$$

where $\rho_1, \ldots, \rho_r$ are real numbers. If for $r = 2s-1$, $J^*_{n,r}$ is the matrix defined by

$$J^*_{n,2s-1} = \begin{bmatrix} \bigcirc & & & & & \bigcirc \\ & 0 & \sigma_1 & \tau_1 & & \\ & & \tau_1 & \sigma_2 & \tau_2 & \\ & & & \ddots & \ddots & \ddots \\ & & & \tau_{s-2} & \sigma_{s-1} & \tau_{s-1} \\ \bigcirc & & & & \tau_{s-1} & \sigma_s \end{bmatrix},$$

where τ_k and σ_k are real numbers, and if for $r = 2s-2$, $J^*_{n,2s-2}$ is the matrix defined as $J_{n,2s-1}$ with $\sigma_1 = 0$, then it is easy to see, using the three-term relation (1.1.1), that the polynomial

$$q^*_{n,r} = \gamma_n \det(xI_n - J_n - J^*_{n,r}) \tag{1.1.4}$$

is a quasi-orthogonal polynomial in the form of (1.1.3), where $\rho_1, \ldots, \rho_r$ are nonlinear functions of τ_k and σ_k in general. Although not every quasi-orthogonal polynomial admits such a symmetric matrix representation, those which do are tied to positive quadrature formulae. We say that $q^*_{n,r}$ admits a positive symmetric matrix representation, if it can be written in the form of (1.1.4) where $J_n + J^*_{n,r}$ has positive subdiagonal elements. It is proved in [43] that a positive quadrature formula is of type $(2n-r-1, n)$, $r \geq 0$, if and only if its nodes are the zeros of a quasi-orthogonal polynomial $q^*_{n,r}$ of order r which admits a positive symmetric matrix representation.

1.2. Background to the theory in several variables

We shall give a quick tour of what is known on the zeros of orthogonal polynomials in several variables and its relation to *cubature* formulae. The word cubature is used for the higher dimensional quadrature which seems to be well accepted among numerical analysts, we shall adopt this terminology throughout this monograph. We stick with the basic notation and facts about orthogonal polynomials in several variables, a more detailed account will be given in Section 2.

We use the standard multiindex notation. Let $\mathbb{N}_0$ be the set of nonnegative integers. For $\alpha = (\alpha_1, \ldots, \alpha_d) \in \mathbb{N}_0^d$ we write $|\alpha| = \alpha_1 + \cdots + \alpha_d$; for

$\mathbf{x} = (x_1, \ldots, x_d) \in \mathbb{R}^d$ we write $\mathbf{x}^\alpha = x_1^{\alpha_1} \cdots x_d^{\alpha_d}$. Let Π^d be the set of polynomials in d variables, and let Π_n^d be the subset of polynomials of total degree at most n; *i.e.*,

$$\Pi_n^d = \left\{ \sum_{|\alpha| \le n} a_\alpha \mathbf{x}^\alpha : \quad a_\alpha \in \mathbb{R}, \quad \mathbf{x} \in \mathbb{R}^d \right\}.$$

We denote by r_n^d the number of linearly independent polynomials of degree exactly n which is equal to the cardinality of the set $\{\alpha \in \mathbb{N}_0^d : |\alpha| = n\}$; it follows that

$$\dim \Pi_n^d = \binom{n+d}{d} \quad \text{and} \quad r_n^d = \binom{n+d-1}{n}.$$

A linear functional $\mathcal{L}$ is called square positive if $\mathcal{L}(p^2) > 0$ for all $p \in \Pi^d$, $p \ne 0$. Examples include all linear functionals expressible as integral against a nonnegative weight function with finite moments; *i.e.*, $\mathcal{L}f = \int f(\mathbf{x})W(\mathbf{x})d\mathbf{x}$. For such a functional $\mathcal{L}$, we apply the Gram-Schmidt orthogonalization process on monomials $\{\mathbf{x}^\alpha\}_{|\alpha|=0}^\infty$ and denote by $\{P_j^n\}_{j=1}^{r_n^d}{}_{n=0}^{\infty}$ a sequence of orthonormal polynomials with respect to $\mathcal{L}$, where the superscript n means that $P_j^n \in \Pi_n^d$ and where the elements are arranged according to the lexicographical order of $\{\alpha \in \mathbb{N}_0^d : |\alpha| = n\}$. Introducing the vector notation

$$\text{(1.2.1)} \qquad \mathbf{P}_n(\mathbf{x}) = \left[P_1^n(\mathbf{x}), P_2^n(\mathbf{x}), \ldots, P_{r_n^d}^n(\mathbf{x})\right]^T,$$

the orthonormal property of $\{P_k^n\}$ can be described as

$$\text{(1.2.2)} \qquad \mathcal{L}(\mathbf{P}_n \mathbf{P}_m^T) = \delta_{m,n} I_{r_n^d},$$

where I_n denotes the $n \times n$ identity matrix. Using this vector notation, it follows that $\mathbf{P}_n$ satisfies a three-term relation ([34, 37])

$$\text{(1.2.3)} \qquad x_i \mathbf{P}_n = A_{n,i}\mathbf{P}_{n+1} + B_{n,i}\mathbf{P}_n + A_{n-1,i}^T \mathbf{P}_{n-1}, \qquad 1 \le i \le d,$$

where $A_{n,i} : r_n^d \times r_{n+1}^d$ and $B_{n,i} : r_n^d \times r_n^d$ are matrices. Throughout this monograph, the notation $A : i \times j$ means that A is a matrix of size $i \times j$.

This vector notation was introduced only recently, but it turned out to be the right notation for the study of orthogonal polynomials in several variables *in general*, in contrast to various special systems. Starting with the three-term relation, many properties of orthogonal polynomials in several variables have been derived (cf. [34–41]). In particular, we can use the coefficient matrices of the three-term relation to define block Jacobi matrices, which are infinite symmetric block tri-diagonal matrices with $B_{n,i}$ on the main diagonal, $A_{n,i}$ and their transpose on the subdiagonals; these matrices can be studied as a commuting family of self-adjoint operators defined on ℓ^2 (cf. [36, 37]). Moreover, the truncated matrices, denoted by $J_{n,i}$ which has n block rows and n block columns, are given as follows,

$$J_{n,i} = \begin{bmatrix} B_{0,i} & A_{0,i} & & & \bigcirc \\ A_{0,i}^T & B_{1,i} & A_{1,i} & & \\ & \ddots & \ddots & \ddots & \\ & & A_{n-3,i}^T & B_{n-2,i} & A_{n-2,i} \\ \bigcirc & & & A_{n-2,i}^T & B_{n-1,i} \end{bmatrix}, \qquad 1 \le i \le d. \tag{1.2.4}$$

Although one cannot expect a representation like (1.1.2) in several variables, $J_{n,i}$ can be used to study the common zeros of orthogonal polynomials in several variables. By a common zero of $\mathbb{P}_n$, say Λ, we mean a zero for every component of $\mathbb{P}_n$; *i.e.*, $P_j^n(\Lambda) = 0$ for all $1 \le j \le r_n^d$. For simplicity, we shall call a common zero of $\mathbb{P}_n$ simply a zero of $\mathbb{P}_n$. The following theorem is proved in [38].

Theorem 1.1. *(i) The point $\Lambda = (\lambda_1, \ldots, \lambda_d)$ is a zero of $\mathbb{P}_n$ if, and only if, it is a joint eigenvalue of $J_n = (J_{n,1}, \ldots, J_{n,d})$; i.e., there is an $\mathbf{x} \ne 0$ such that*

$$J_{n,i}\mathbf{x} = \lambda_i \mathbf{x}, \quad 1 \le i \le d; \tag{1.2.5}$$

moreover, a joint eigenvector of Λ is $(\mathbb{P}_0^T(\Lambda), \ldots, \mathbb{P}_{n-1}^T(\Lambda))^T$.
(ii) The orthogonal polynomial $\mathbb{P}_n$ has $N = \dim \Pi_{n-1}^d$ distinct real zeros if, and only if,

$$A_{n-1,i}A_{n-1,j}^T = A_{n-1,j}A_{n-1,i}^T, \quad 1 \le i, j \le d. \tag{1.2.6}$$

For $d = 1$ the second part of the theorem is trivial; every orthogonal polynomial of degree n has $n = \dim \Pi_{n-1}^1$ distinct real zeros. For $d \ge 2$, however, it is clear from (1.2.6) that not every polynomial vector $\mathbf{P}_n$ has $\dim \Pi^d_{n-1}$ many real zeros (matrix multiplication is not a commuting operation). Actually, polynomials which do have these many zeros are rare, the only known examples where a measure is given such that (1.2.6) holds for all n have been found only very recently in [3, 29]. One important aspect of this lies in its connection to cubature.

For a square positive $\mathcal{L}$, a cubature formula of degree $2n-1$ is a linear functional

$$(1.2.7) \qquad \mathcal{I}_n(f) = \sum_{k=1}^{N} \lambda_k f(\mathbf{x}_k)\,, \quad \lambda_k > 0, \quad \mathbf{x}_k \in \mathbb{R}^d,$$

defined on Π^d, such that

$$(1.2.8) \qquad \mathcal{L}(f) = \mathcal{I}_n(f), \qquad \forall f \in \Pi^d_{2n-1}.$$

The points $\mathbf{x}_1, \ldots, \mathbf{x}_N$ are called nodes and $\lambda_1, \ldots, \lambda_N$ weights. Such a formula is called minimal, if N, the number of nodes, is minimal among all cubature formulae of degree $2n-1$. It is well-known (cf. [30]) that

$$(1.2.9) \qquad N \geq \dim \Pi^d_{n-1}$$

in general. We shall call formulae attaining this lower bound Gaussian, which, in case they exist, are clearly minimal.

The existence of Gaussian cubature formulae was first considered by Radon [26] in 1948 for $d = 2$ and $n = 3$. The work of Radon was continued by Mysovskikh and his school (cf. [21–24, 15], see further reference in [23]). It was proved by Mysovskikh that a Gaussian cubature formula exists if, and only if, the corresponding orthogonal polynomials of degree n, in our notation $\mathbf{P}_n$, have $\dim \Pi^d_{n-1}$ many distinct real zeros. However, since $\mathbf{P}_n$ does not have these many zeros in general, Gaussian cubature formulae in general do not exist. Nevertheless, it is clear that for any square positive linear functional, a minimal cubature formula exists. In [17–19], Möller studied minimal cubature and showed that there is an improved lower bound for the number of nodes. For $d = 2$ this bound, in our notation, states (cf. Section 5.1 below),

$$(1.2.10) \qquad N \geq \dim \Pi^2_{n-1} + \frac{1}{2}\operatorname{rank}(A_{n-1,1}A^T_{n-1,2} - A_{n-1,2}A^T_{n-1,1}).$$

Note that if Gaussian cubature exists, thus (1.2.6) holds, then this lower bound coincides with (1.2.9). It is evident from Möller's work that the nodes of a minimal cubature formula attaining his improved lower bound are common zeros of certain orthogonal polynomials of degree n.

Möller used algebraic ideal theory in his study of cubature, or equivalently, the study of common zeros of polynomials, which led to a major step forward. Although there have been quite a few papers following the work of Mysovskikh and Möller, progress has been slow and concentrated more or less around Möller's bound. This is, at least in part, due to the fact that the structure of the common zeros of orthogonal polynomials has not been completely understood through algebraic ideal theory, a subject which is still under development itself. Moreover, the picture becomes even gloomier in the case when Möller's bound (1.2.10) is not attained, for then the approach via algebraic ideal theory does not seem to apply and no other approach has been known that can be used to attack the problem.

The purpose of this monograph is to introduce a new approach and use it to conduct a systematic study of the common zeros of sets of polynomials in several variables and of the corresponding cubature formulae. This approach is based on our recent work on orthogonal polynomials in several variables; it does not use algebraic ideal theory. We provide a complete characterization of the common zeros of certain quasi-orthogonal polynomials and of the related cubature formulae. Moreover, the results can be applied to all such formulae in contrast to the previous approaches which deal primarily with those ones which attain the lower bounds (1.2.10).

1.3. Outline of the content

The monograph is divided into several parts outlined in this section. In Section 2, we introduce notations and preliminaries concerning orthogonal polynomials in several variables, and prove several lemmas of general nature which are of interests in themselves. In section 3, we present two preliminary results which motivated our study; the second one gives the necessary conditions for the existence of minimal cubature formulae and defines the set of polynomials whose common zeros we will study.

Since in the most general setting, which amounts to study common zeros of quasi-orthogonal polynomials of various degrees, the situation is very complicated, we devote Section 4 to the study of the first nontrivial case beyond Theorem 1.1: common zeros of the set of polynomials $\{U^T\mathbf{P}_n, \mathbf{Q}_{n+1}\}$, where for a positive integer $\sigma \le r_n^d$, $U : r_n^d \times r_n^d - \sigma$, is a matrix with rank $r_n^d - \sigma$, and $\mathbf{Q}_{n+1} = \mathbf{P}_{n+1} + \Gamma_1 \mathbf{P}_n + \Gamma_2 \mathbf{P}_{n-1}$ where Γ_1 and Γ_2 are certain matrices. Our central theorem, Theorem 4.1.4 which consists of a characterization of the common zeros, is in the spirit of Theorem 1.1; the existence of real common zeros is characterized in terms of nonlinear matrix equations satisfied by the matrix U, or the matrix V defined by $U^TV = 0$, and the matrices Γ_1 and Γ_2. To complete the proof of this theorem and to establish the connection between common zeros and cubature formulae, we prove an analogy of the Christoffel-Darboux formula in Section 4.2, which is used to study a special Lagrange interpolation in Section 4.3. Assuming the maximal number of common zeros of $U^T\mathbf{P}_n$ and $\mathbf{Q}_{n+1}$ to exist, we prove, in Section 4.4, the existence of cubature formulae of degree $2n-1$.

The cubature formulae developed in Section 4 may not be minimal ones, although any minimal cubature formula which is based on the common zeros of some $U^T\mathbf{P}_n$ and $\mathbf{Q}_{n+1}$ is included in our characterization. For a given linear functional $\mathcal{L}$ and a cubature formula of degree $2n-1$, the only existing theoretic way of deciding whether the formula is minimal is to check the number of nodes against a known lower bound. In this respect, Möller's lower estimates are still the best ones available. Because of their importance, we devote Section 5 to the results centered around these lower bounds. In Section 5.1, we consider the first lower bound (1.2.9) and prove an interesting necessary condition for it to hold. In Section 5.2 we prove Möller's first lower bound in arbitrary dimensions using our vector-matrix notation. In Section 5.3, we discuss when this bound will be attained in the light of our results obtained in Section 4. In particular, we will show that, if this bound is attained, then it is completely characterized by a special case of results stated in Section 4. There is a second lower bound given by Möller for centrally symmetric weight functions, which offers a stronger lower bound when $d > 2$. In Section 5.4, we offer a slightly different proof of this bound to put it in the light of our general characterization.

In Section 6, we consider various examples to illuminate our results. Since the characterization in Section 4.1 is given in terms of a nonlinear system of matrix equations which are, however, very difficult to solve in general, our examples are given only for product weight functions. We discuss in general the problem of solving the matrix equations in Section 6.1. The easiest examples among all product weight functions seem to be the product Chebyshev weights of the first and the second kind, respectively, which are used to construct various examples in Section 6.2 to illustrate the theorems. In Section 6.3, we discuss solving the matrix equations for general product weights and small order n.

In Section 7, we present the characterization of the common zeros of sets of polynomials and of cubature formulae in the most general setting. The results are fairly general; they provide, in particular, a way of describing cubature formulae of degree $2n-1$, which should open the way for further characterizations, possibly more accessible. We will follow the approach in the simplest case in Section 4; however, there will be little overlap with the proofs in Section 4.

The method developed in the previous sections is also useful for deriving cubature formulae of degree $2n-2$. These formulae are based on zeros of quasi-orthogonal polynomials of odd order, and have been studied in the literature almost as a separate topic (cf. [20, 27, 40]). However, as we shall show in Section 8 that the characterization of the common zeros of the corresponding polynomials is more or less contained in our results in Sections 4 and 7. Although minimal cubature formulae of degree $2n-2$ are considered to be very different from those of degree $2n-1$, which is perhaps true as far as the lower bound is concerned, our results indicate that there is hardly a significant difference of one versus the other from a theoretic viewpoint. In Section 8.1 we present the preliminaries; the new characterization is given in Section 8.2 which goes far beyond the existing results. Using this new characterization, in Section 8.3 we construct a cubature formula of degree $2n-2$ for product Chebyshev weight of the first kind, which is a weight function for which the early approaches allegedly failed to yield any solution.

From our development it is evident that there is no reason that one should stop studying cubature formulae at degree $2n-2$. Actually, one could consider formulae of degree $2n-s$ for $s \geq 1$ just as in the case of one variable; it is just that the situation becomes even more complicated. Nevertheless, in Section 9.1, we will briefly discuss cubature formulae of degree $2n-s$ in general, and provide one example to explain the situation. Section 9.2 serves as conclusion to the entire monograph, including remarks about aspects of the construction of cubature formulae in general.

2. Preliminaries and Lemmas

This section contains three subsections. The first one summaries the theorems on orthogonal polynomials in several variables, the second one contains a characterization of centrally symmetric linear functionals, and the third one collects various lemmas.

2.1. Orthogonal polynomials in several variables

2.1.1. General theory. For earlier results on orthogonal polynomials in several variables we refer to [7] for results before 1953, to [8, 23] for more recent results, and to [31] for orthogonal polynomials in two variables. One should mention that there have been results by many authors concerning special systems of orthogonal polynomials in several variables, such as product Hermite polynomials, Jacobi polynomials on a cube, on a simplex, and on a ball, just to name a few; some of the systems have important applications in connection with group representation and other mathematical branches. In contrast to the study of special systems, we are mainly concerned with the general theory; *i.e.*, properties shared by every system of orthogonal polynomials. In this respect, the vector notation (1.2.1), which we introduced in [34], proves to be very fruitful; it allows us to develop a theory of orthogonal polynomials in several variables which in many ways parallels the theory in one variable. In the following we summarize the essential parts of the theory developed in [34–41] which will be used in the monograph.

Throughout the monograph let $\mathcal{L}$ be a square positive linear functional. For convenience we always assume $\mathcal{L}(1) = 1$ so that $\mathbb{P}_0 = 1$.

For $n \in \mathbb{N}_0$ we write $\mathbf{x}^n = (\mathbf{x}^\alpha)_{|\alpha|=n}$, which is a vector in $\mathbb{R}^{r_n^d}$ whose elements are arranged according to the lexicographical order of $\{\alpha \in \mathbb{N}_0^d : |\alpha| = n\}$. The orthogonal polynomials $\mathbb{P}_n$ of (1.2.1) can be written as

$$\mathbb{P}_n = G_n\mathbf{x}^n + G_{n,n-1}\mathbf{x}^{n-1} + G_{n,n-2}\mathbf{x}^{n-2} + \dots \tag{2.1.1}$$

where $G_{n,i} : r_n^d \times r_{n-i}^d$ are matrices. We call $G_n = G_{n,n}$ the leading coefficient of $\mathbf{P}_n$. It is proved in [39] that G_n is invertible.

Our development of the theory of orthogonal polynomials in several variables is based on the three-term relation (1.2.3), which we restate as

Three-term relation. *For $n \geq 0$, there exist matrices $A_{n,i} : r_n^d \times r_{n+1}^d$ and $B_{n,i} : r_n^d \times r_n^d$, such that*

$$x_i \mathbf{P}_n = A_{n,i}\mathbf{P}_{n+1} + B_{n,i}\mathbf{P}_n + A_{n-1,i}^T \mathbf{P}_{n-1}, \qquad 1 \leq i \leq d, \tag{2.1.2}$$

where we define $\mathbf{P}_{-1} = 0$ and $A_{-1,i} = 0$.

This relation should be compared to (1.1.1). The condition in place of $a_n \neq 0$ in (1.1.1) turns out to be

Rank conditions. *For $n \geq 0$, $\operatorname{rank} A_{n,i} = r_n^d$ for $1 \leq i \leq d$, and*

$$\operatorname{rank} A_n = r_{n+1}^d, \qquad A_n = (A_{n,1}^T | \ldots | A_{n,d}^T)^T. \tag{2.1.3}$$

The importance of these notions is justified by the following fundamental theorem which is the analogue of Favard's theorem in one variable ([12, 34, 36]).

Favard's theorem. *Let $\{\mathbf{P}_n\}_{n=0}^{\infty}$, $\mathbf{P}_0 = 1$, be a sequence in Π^d. Then the following statements are equivalent:*

1) There exists a linear functional which is square positive and makes $\{\mathbf{P}_n\}_{n=0}^{\infty}$ an orthonormal basis in Π^d;

2) For $n \geq 0$, $1 \leq i \leq d$, there exist matrices $A_{n,i} : r_n^d \times r_{n+1}^d$ and $B_{n,i} : r_n^d \times r_n^d$ such that the polynomial vectors $\mathbf{P}_n$ satisfy the three-term relation (2.1.2) and the rank condition (2.1.3).

When $\mathbf{P}_n$ is orthonormal with respect to $\mathcal{L}$, the matrices in the three-term relation are expressible as

$$A_{n,i} = \mathcal{L}(x_i \mathbf{P}_n \mathbf{P}_{n+1}^T) \quad \text{and} \quad B_{n,i} = \mathcal{L}(x_i \mathbf{P}_n \mathbf{P}_n^T). \tag{2.1.4}$$

Another important set of conditions satisfied by the orthonormal polynomials $\mathbf{P}_n$ are the following:

Commuting conditions. For $n \ge 0$, $1 \le i, j \le d$,

$$A_{n,i}A_{n+1,j} = A_{n,j}A_{n+1,i}\ , \tag{2.1.5}$$

$$A_{n,i}B_{n+1,j} + B_{n,i}A_{n,j} = B_{n,j}A_{n,i} + A_{n,j}B_{n+1,i}\ , \tag{2.1.6}$$

$$A^T_{n-1,i}A_{n-1,j} + B_{n,i}B_{n,j} + A_{n,i}A^T_{n,j} = A^T_{n-1,j}A_{n-1,i} + B_{n,j}B_{n,i} + A_{n,j}A^T_{n,i}\ . \tag{2.1.7}$$

We call these conditions commuting because they are the ones that make the linear operators defined by the block Jacobi matrices formally commuting. The spectral theorem for a commuting family of self-adjoint operators has been used to study the integral representation of the linear functional in Favard's theorem (see [36, 37]). These conditions are especially important in the proof of Theorem 1.1 and in our study.

The rank condition (2.1.3) implies that there exists a matrix $D^T_n : r^d_{n+1} \times dr^d_n$, which we write as $D^T_n = (D^T_{n,1}|\dots|D^T_{n,d})$ where $D^T_{n,i} : r^d_{n+1} \times r^d_n$, which satisfies

$$D^T_n A_n = \sum_{i=1}^{d} D^T_{k,i}A_{k,i} = I_{r^d_{n+1}}. \tag{2.1.8}$$

It is important to note that D^T_n is not unique, see **2.3.1.** For each D^T_n, we can derive from (2.1.2) the following:

Recursive relation. For $n \ge 0$

$$\mathbf{P}_{n+1}(\mathbf{x}) = \sum_{i=1}^{d} D^T_{n,i}x_i\mathbf{P}_n(\mathbf{x}) + E_n\mathbf{P}_n(\mathbf{x}) + F_n\mathbf{P}_{n-1}(\mathbf{x}) \tag{2.1.9}$$

where $\mathbf{P}_{-1} = 0$ *and where*

$$E_n = -\sum_{i=1}^{d} D^T_{n,i}B_{n,i} \quad \text{and} \quad F_n = -\sum_{i=1}^{d} D^T_{n,i}A^T_{n-1,i}. \tag{2.1.10}$$

The reproducing kernel function of the orthonormal system of polynomials $\{\mathbf{P}_n\}_{n=0}^{\infty}$, denoted by $\mathbf{K}_n(\cdot,\cdot)$, is defined as

$$\mathbf{K}_n(\mathbf{x},\mathbf{y}) = \sum_{k=0}^{n-1} \mathbf{P}^T_k(\mathbf{x})\mathbf{P}_k(\mathbf{y});$$

using (2.1.2) it is easy to show that $\mathbf{K}_n(\mathbf{x},\mathbf{y})$ satisfies the following ([34]):

Christoffel-Darboux formula. *For $n \geq 1$, $1 \leq i \leq d$,*

$$(2.1.11) \qquad \mathbf{K}_n(\mathbf{x},\mathbf{y}) = \frac{[A_{n-1,i}\mathbf{P}_n(\mathbf{x})]^T\mathbf{P}_{n-1}(\mathbf{y}) - \mathbf{P}_{n-1}^T(\mathbf{x})[A_{n-1,i}\mathbf{P}_n(\mathbf{y})]}{x_i - y_i}.$$

This formula is very useful in studying the properties of the zeros of $\mathbf{P}_n$. Again, by a zero of $\mathbf{P}_n$ we mean a common zero of the components of $\mathbf{P}_n$. If Λ is a zero of $\mathbf{P}_n$ and if at least one partial derivative of a component of $\mathbf{P}_n$ is not zero at Λ, then we call Λ a *simple* zero. We have ([23, 38])

Properties of Zeros. *All zeros of $\mathbf{P}_n$ are real and simple. Two consecutive polynomials $\mathbf{P}_n$ and $\mathbf{P}_{n-1}$ do not have common zeros.*

Further results on the general theory of orthogonal polynomials in several variables can be found in [11, 12, 23, 34–41].

2.1.2. Examples. To explain our vector-matrix notation we give several examples of bivariate orthogonal polynomials. For the case of two variables we take the subindex k of the orthogonal polynomials $\{P_k^n\}$ to run from 0 to n instead of from 1 to $r_n^2 = n+1$. In particular, our vector notation (1.2.1) takes the form $\mathbf{P}_n = (P_0^n, \ldots, P_n^n)^T$.

Example 2.1. *Product orthogonal polynomials.* Let p_n be the orthonormal polynomials with respect to a measure μ which satisfy the three-term relation (1.1.1). Let σ be the product measure on $\mathbf{R}^2$ defined by $d\sigma(x,y) = d\mu(x)d\mu(y)$. It is easy to see that the orthonormal polynomials in two variables with respect to σ are given by

$$P_k^n(x,y) = p_{n-k}(x)p_k(y), \quad 0 \leq k \leq n.$$

From (2.1.4) it is easy to verify that the coefficient matrices of the three-term relation (2.1.2) are of the following form:

$$(2.1.12) \qquad A_{n,1} = \begin{bmatrix} a_n & & \bigcirc & 0 \\ & \ddots & & \vdots \\ \bigcirc & & a_0 & 0 \end{bmatrix} \quad \text{and} \quad A_{n,2} = \begin{bmatrix} 0 & a_0 & & \bigcirc \\ \vdots & & \ddots & \\ 0 & \bigcirc & & a_n \end{bmatrix};$$

and

$$
(2.1.13) \qquad B_{n,1} = \begin{bmatrix} b_n & & \bigcirc \\ & \ddots & \\ \bigcirc & & b_0 \end{bmatrix} \quad \text{and} \quad B_{n,2} = \begin{bmatrix} b_0 & & \bigcirc \\ & \ddots & \\ \bigcirc & & b_n \end{bmatrix}.
$$

Example 2.2. In [13], Koornwinder gave a general method of generating orthogonal polynomials of two variables from orthogonal polynomials of one variable. We restrict ourselves to two special cases.

Let ν be a nondecreasing function on (c, d) with finite moments and let $\{q_n\}_{n=0}^{\infty}$ be the orthonormal polynomials with respect to $d\nu$. We write the three-term relation satisfied by q_n as

$$
xq_n = a_n(\nu)q_{n+1} + b_n(\nu)q_n + a_{n-1}(\nu)q_{n-1}, \quad n \geq 0,
$$

where $q_0 = 1$ and $q_{-1} = 0$. Let μ be a nondecreasing function on the interval (a, b) with finite moments and let ρ be a positive function on (a, b). For each $k \in \mathbb{N}_0$ let the polynomials $\{p_{n,k}\}_{n=0}^{\infty}$ be orthonormal with respect to the measure $\rho^{2k+1}d\mu$. We denote the leading coefficient of $p_{n,k}$ by $\gamma_{n,k}$ and write the three-term relation for $p_{n,k}$ as

$$
xp_{n,k} = a_{n,k}p_{n+1,k} + b_{n,k}p_{n,k} + a_{n-1,k}p_{n-1,k}, \qquad n \geq 0,
$$

where $p_{0,k} = 1$ and $p_{-1,k} = 0$, $k \geq 0$. We consider,

Case I. ρ is a linear polynomial: $\rho(x) = 1 - x$, $(a, b) = (0, 1)$,

Case II. ρ is the square root of a quadratic polynomial: $\rho(x) = (1 - x^2)^{1/2}$, and both μ' and ν' are even functions on $(-1, 1)$.

For these two cases, we define polynomials P_k^n of two variables by

$$
(2.1.14) \qquad P_k^n(x, y) = p_{n-k,k}(x)(\rho(x))^k q_k\Big(\frac{y}{\rho(x)}\Big), \quad 0 \leq k \leq n.
$$

It can be easily verified that $P_k^n \in \Pi_n^2$; moreover, $\{P_k^n\}$ are orthonormal with respect to the measure

$$
d\sigma = \mu'(x)\nu'(\rho^{-1}(x)y)\,dxdy, \quad \text{on} \quad R = \{(x, y) : a < x < b, \ c\rho(x) < y < d\rho(x)\},
$$

that is

$$\iint_R P_k^n(x,y)P_j^m(x,y)d\sigma = \delta_{m,n}\delta_{k,j}.$$

Actually this construction works as long as ρ is either a linear function or a square root of a nonnegative quadratic polynomial, we refer to [13] for the general discussion. Because of the geometric structure of R, the two cases are termed, respectively, orthogonal polynomials on a triangular region, and orthogonal polynomials on a circular region. Using (2.1.4), we can then determine the matrices $A_{n,i}$ and $B_{n,i}$, $i = 1,2$, in the three-term relation (2.1.2) for these orthogonal polynomials in two variables. The results are as follows:

Orthogonal polynomials on the triangular region:

$$d\sigma = \mu'(x)\nu'(\frac{y}{1-x}), \quad \text{on} \quad T = \{(x,y) : 0 < x, y < 1, \ x + y < 1\};$$

The coefficient matrices are given by:

$$A_{n,1} = \begin{bmatrix} a_{n,0} & & & \bigcirc & 0 \\ & a_{n-1,1} & & & 0 \\ & & \ddots & & \vdots \\ \bigcirc & & & a_{0,n} & 0 \end{bmatrix}, \quad B_{n,1} = \begin{bmatrix} b_{n,0} & & & \bigcirc \\ & b_{n-1,1} & & \\ & & \ddots & \\ \bigcirc & & & b_{0,n} \end{bmatrix},$$

and

$$A_{n,2} = \begin{bmatrix} e_{0,n} & d_{0,n} & & & \bigcirc & 0 \\ c_{1,n} & e_{1,n} & d_{1,n} & & & \vdots \\ & \ddots & \ddots & \ddots & & \vdots \\ & & c_{n-1,n} & & d_{n-1,n} & 0 \\ \bigcirc & & & c_{n,n} & e_{n,n} & d_{n,n} \end{bmatrix},$$

where

$$c_{k,n} = a_{k-1}(\nu)\frac{\gamma_{n-k,k}}{\gamma_{n-k+2,k-1}}, \quad d_{k,n} = a_k(\nu)\frac{\gamma_{n-k,k}}{\gamma_{n-k,k+1}},$$

and

$$e_{k,n} = b_k(\nu)\int_a^b p_{n-k,k}(x)p_{n+1-k,k}(x)(\rho(x))^{2k+2}d\mu,$$

and, finally,

$$B_{n,2} = \begin{bmatrix} b_0(\nu) & g_{0,n} & & & \bigcirc \\ g_{0,n} & b_1(\nu) & g_{1,n} & & \\ & \ddots & \ddots & \ddots & \\ & & g_{n-2,n} & b_{n-1}(\nu) & g_{n-1,n} \\ \bigcirc & & & g_{n-1,n} & b_n(\nu) \end{bmatrix},$$

where

$$g_{k,n} = a_k(\nu) \int_a^b p_{n-k,k}(x) p_{n-k-1,k+1}(x) (\rho(x))^{2k+3} d\mu .$$

If $\mu'(x) = x^\alpha(1-x)^{\beta+\gamma}$ and $\nu'(x) = x^\beta(1-x)^\gamma$, $\alpha, \beta, \gamma > -1$, then P_k^n are the classical Jacobi polynomials on the triangle T which are orthogonal with respect to the weight function $x^\alpha y^\beta (1-x-y)^\gamma$.

Orthogonal polynomials on circular region:

$$d\sigma = \mu'(x)\nu'(y(1-x^2)^{-1/2}), \quad \text{on} \quad D = \{(x,y) : x^2+y^2 < 1\};$$

Here the coefficient matrices are given as follows: $B_{n,1} = B_{n,2} = 0$, $A_{n,1}$ and $A_{n,2}$ are of the same form as in the triangular case except that now

$$c_{k,n} = -a_{k-1}(\nu)\frac{\gamma_{n-k,k}}{\gamma_{n-k+2,k-1}}, \qquad e_{k,n} = 0.$$

If $\mu'(x) = (1-x^2)^\alpha$ and $\nu'(x) = (1-x^2)^\alpha$, $\alpha > -1$, then P_k^n are the classical Jacobi polynomials on the unit disk D which are orthogonal with respect to the weight function $(1-x^2-y^2)^\alpha$.

Example 2.3. Let μ be a nondecreasing function on $\mathbb{R}$ with finite moments. Let $\{p_n\}_{n=0}^\infty$ be orthonormal polynomials with respect to μ; it satisfies the three-term relation (1.1.1). For a given $n \in \mathbb{N}_0$, let

$$u = x + y, \qquad v = xy$$

and define

$$P_k^{n,(-\frac{1}{2})}(u,v) = \begin{cases} p_n(x)p_k(y) + p_n(y)p_k(x), & \text{if } k < n, \\ \sqrt{2}p_n(x)p_n(y), & \text{if } k = n, \end{cases} \tag{2.1.15}$$

and

$$P_k^{n,(\frac{1}{2})}(u,v) = \frac{p_{n+1}(x)p_k(y) - p_{n+1}(y)p_k(x)}{x-y}. \tag{2.1.16}$$

Then $P_k^{n,(\pm\frac{1}{2})}$ are polynomials of total degree n in the variables u and v. Moreover, let $W(u,v) = \mu'(x)\mu'(y)$; it is not hard to verify that $\{P_k^{n,(\pm\frac{1}{2})}\}$ are orthonormal polynomials in two variables with respect to the measures

$$d\sigma = (u^2 - 4v)^{\pm 1/2} W(u,v)dudv, \quad \text{on} \quad R = \{(u,v)|(x,y) \in \mathbb{R} \times \mathbb{R}, \text{ and } x < y\},$$

respectively (cf. [13, p. 468]). We note that the support set of $d\sigma$ is the image of a triangular-like region under the nonlinear transformation $u = x + y$ and $v = xy$. For example, if the support set of μ is $[-1,1]$, then the support set of σ is bounded by the lines $u = v + 1$, $u = -v - 1$, and the parabola $u^2 = 4v$.

We denote by $A_{n,i}^{(\pm\frac{1}{2})}$ and $B_{n,i}^{(\pm\frac{1}{2})}$ the coefficient matrices of the three-term relation for $\mathbf{P}_n^{(\pm\frac{1}{2})}$, respectively. Then using the definition of $P_k^{n,(-\frac{1}{2})}$ and the three-term relation satisfied by p_n as given in (1.1.1), it is not hard to derive (cf. [29]) that

$$A_{n,1}^{(-\frac{1}{2})} = a_n \begin{bmatrix} 1 & & & \bigcirc & 0 \\ & \ddots & & & \vdots \\ & & 1 & & 0 \\ \bigcirc & & & \sqrt{2} & 0 \end{bmatrix},$$

$$B_{n,1}^{(-\frac{1}{2})} = \begin{bmatrix} b_0 & a_0 & & & \bigcirc \\ a_0 & b_1 & a_1 & & \\ & \ddots & \ddots & \ddots & \\ & & a_{n-2} & b_{n-1} & \sqrt{2}a_{n-1} \\ \bigcirc & & & \sqrt{2}a_{n-1} & b_n \end{bmatrix} + b_n I_{n+1},$$

and

$$A_{n,2}^{(-\frac{1}{2})} = a_n \begin{bmatrix} b_0 & a_0 & & & \bigcirc & 0 \\ a_0 & b_1 & a_1 & & & \vdots \\ & \ddots & \ddots & \ddots & & \vdots \\ & & a_{n-2} & b_{n-1} & \sqrt{2}a_{n-1} & 0 \\ \bigcirc & & & \sqrt{2}a_{n-1} & \sqrt{2}b_n & a_n \end{bmatrix}$$

$$B_{n,2}^{(-\frac{1}{2})} = \begin{bmatrix} b_0 & a_0 & & & \bigcirc \\ a_0 & b_1 & a_1 & & \\ & \ddots & \ddots & \ddots & \\ & & a_{n-2} & b_{n-1} & \sqrt{2}a_{n-1} \\ \bigcirc & & & \sqrt{2}a_{n-1} & b_n \end{bmatrix} + a_{n-1}^2 \begin{bmatrix} \bigcirc & & \\ & 1 & 0 \\ & 0 & 0 \end{bmatrix}.$$

The coefficient matrices $A_{n,1}^{(\frac{1}{2})}$ and $B_{n,1}^{(\frac{1}{2})}$ can be derived similarly, we refer to [29].

Using symmetric polynomials the above definitions can be extended to $\mathbb{R}^d$ in a natural way, which we will discuss in Section 9.

2.1.3. Quasi-orthogonal polynomials. Let $r, n \in \mathbb{N}_0$ and $r \leq n$. If $Q \in \Pi_n^d$ is orthogonal to polynomials of degree less than or equal to $n - r - 1$, we call Q a quasi-orthogonal polynomials of order r. Since every polynomial of degree n can be written in terms of the orthonormal basis $\{\mathbf{P}_k, 0 \leq k \leq n\}$ of Π_n^d, it follows from the definition that the components of

$$\mathbf{Q}_{n,r} = \mathbf{P}_n + \Gamma_{1,r}\mathbf{P}_{n-1} + \ldots + \Gamma_{r,r}\mathbf{P}_{n-r}, \qquad \Gamma_{j,r} : r_n^d \times r_{n-j}^d,$$

are quasi-orthogonal polynomials of degree n and order r which form a basis for all such polynomials.

In the past only quasi-orthogonal polynomials of order 1; *i.e.*, those of the form $\mathbf{Q}_n = \mathbf{P}_n + \Gamma\mathbf{P}_{n-1}$, have been studied. Their zeros are associated with cubature formulae of even degree (cf. [20, 27, 37, 40] and Section 8). However, it turns out that the general quasi-orthogonal polynomials play a decisive role in cubature both for formulae of odd and even degrees.

2.2. Centrally symmetric linear functional

Let $\mathcal{L}$ be a square positive linear functional. Of special interests are examples of $\mathcal{L}$ expressible as integrals against a nonnegative weight function with finite moments; *i.e.*, $\mathcal{L}f = \int f(\mathbf{x})W(\mathbf{x})d\mathbf{x}$; in such cases we shall refer to the weight W instead of to the linear functional $\mathcal{L}$. Let $\Omega \subset \mathbb{R}^d$ be the support set of W. The weight W is said to be *centrally symmetric*, if

$$\text{(2.2.1)} \qquad \mathbf{x} \in \Omega \Rightarrow -\mathbf{x} \in \Omega, \quad \text{and} \quad W(\mathbf{x}) = W(-\mathbf{x}).$$

We also call the linear functional $\mathcal{L}$ to be centrally symmetric, if it satisfies

$$\mathcal{L}(\mathbf{x}^\alpha) = 0, \quad \alpha \in \mathbf{N}^d, \quad |\alpha| = \text{odd integer}. \tag{2.2.2}$$

It is easily seen that when $\mathcal{L}$ is expressible as an integral against W, (2.2.2) follow from (2.2.1). As examples we mention the product weight functions in Exampl 2.1 which are centrally symmetric if μ' is even on a symmetric interval; the weigh functions on a circular region in Example 2.2 are all centrally symmetric since w take both μ' and ν' to be even on $(-1, 1)$; however, weight functions on a triangula region in Example 2.2 and those in Example 2.3 are not centrally symmetric.

A central symmetric $\mathcal{L}$ has a relatively simple structure which allows us to get more information about the structure of orthogonal polynomials. The followin theorem connects the central symmetry of $\mathcal{L}$ to the coefficient matrices in the three term relation.

Theorem 2.2.1. *Let $\mathcal{L}$ be a square positive linear functional, and let $B_{n,i}$ be the coefficient matrices in (2.1.2) for the orthogonal polynomials with respect to $\mathcal{L}$. Then, $\mathcal{L}$ is centrally symmetric if, and only if, $B_{n,i} = 0$ for all $n \in \mathbf{N}_0$ and $1 \le i \le d$.*

Proof. First we assume that $\mathcal{L}$ is centrally symmetric. From (2.1.4), we have

$$B_{0,i} = \mathcal{L}(x_i \mathbf{P}_0 \mathbf{P}_0^T) = \mathcal{L}(x_i) = 0.$$

Therefore, from (2.1.9), $\mathbf{P}_1 = \sum x_i D_{0,i}^T$ which implies that the constant terms in the components of $\mathbf{P}_1$ vanish. Since

$$\mathbf{P}_n = \sum_{i=1}^{d} x_i D_{n-1,i}^T \mathbf{P}_{n-1} + F_{n-1}\mathbf{P}_{n-2}$$

for $B_{n-1,i} = 0$, $1 \le i \le d$, we can use induction to conclude that if $B_{k,i} = 0$ for $0 \le k \le n$, then the components of $\mathbf{P}_n$ are sums of even powers of x_i if n is even, and odd powers of x_i if n is odd. Suppose now we have proved $B_{k,i} = 0$ for $1 \le k \le n-1$. By (2.1.9) and (2.1.10),

$$\begin{aligned} B_{n,i} &= \mathcal{L}(x_i \mathbf{P}_n \mathbf{P}_n^T) \\ &= \sum_{j=0}^{d} \sum_{k=0}^{d} \mathcal{L}(x_i D_{n-1,j}^T (x_j \mathbf{P}_{n-1} + A_{n-2,j}^T \mathbf{P}_{n-2})(x_k \mathbf{P}_{n-1} + A_{n-2,k}^T \mathbf{P}_{n-2})^T D_{n-1,k}). \end{aligned}$$

By induction, the components in $(x_j\mathbf{P}_{n-1} + A_{n-2,j}^T\mathbf{P}_{n-2})(x_k\mathbf{P}_{n-1} + A_{n-2,k}^T\mathbf{P}_{n-2})^T$ are polynomials that are sums of even powers, thus, whose x_i multiples are sums of odd powers. Since $\mathcal{L}$ is centrally symmetric, we obtain $B_{n,i} = 0$.

On the other hand, assuming that $B_{n,i} = 0$ for all $n \in \mathbb{N}_0$, then as we proved above the components of $\mathbf{P}_n$ are sums of even powers of x_i if n is even, and odd powers of x_i if n is odd. From $B_{0,i} = \mathcal{L}(x_i\mathbf{P}_0\mathbf{P}_0^T) = \mathcal{L}(x_i) = 0$, it follows that $\mathcal{L}(x_i) = 0$. We now use induction to prove that

$$\{B_{k,i} = 0, \quad 1 \le k \le n\} \quad \Rightarrow \quad \{\mathcal{L}(x_1^{\alpha_1} \dots x_d^{\alpha}) = 0, \quad \alpha_1 + \dots + \alpha_d = 2n+1\}.$$

Suppose that the claim has been proved for $k \le n-1$. We have by (2.1.2) and (2.1.4)

$$\begin{aligned} A_{n-1,j}B_{n,i}A_{n-1,k}^T =& \mathcal{L}(x_i(x_j\mathbf{P}_{n-1} - A_{n-2,j}^T\mathbf{P}_{n-2})(x_k\mathbf{P}_{n-1} - A_{n-2,k}^T\mathbf{P}_{n-2})^T) \\ =& \mathcal{L}(x_ix_jx_k\mathbf{P}_{n-1}\mathbf{P}_{n-1}^T) - \mathcal{L}(x_ix_j\mathbf{P}_{n-1}\mathbf{P}_{n-2}^T)A_{n-2,k} \\ &- A_{n-2,j}^T\mathcal{L}(x_ix_k\mathbf{P}_{n-1}\mathbf{P}_{n-2}^T) + A_{n-2,j}^TB_{n-2,i}A_{n-2,k}. \end{aligned}$$

By induction, we have that $\mathcal{L}(x_ix_j\mathbf{P}_{n-1}\mathbf{P}_{n-2}^T) = 0$ and $\mathcal{L}(x_ix_k\mathbf{P}_{n-1}\mathbf{P}_{n-2}^T) = 0$, since the polynomial components of these matrices are sums of odd powers and of total degree $\le 2n-1$. Therefore, we have

$$\mathcal{L}(x_ix_jx_k\mathbf{P}_{n-1}\mathbf{P}_{n-1}^T) = 0.$$

Multiplying the above matrix from left and right by A_{k,l_i} and A_{k,l_i}^T, respectively, for $1 \le l_i \le d$ and $i = n-2, \dots, 1$, and using the three-term relation, we can repeat the above argument and derive $\mathcal{L}(x_1^{\alpha_1} \dots x_d^{\alpha_d}) = 0$, $\alpha_1 + \dots + \alpha_d = 2n+1$, in finitely many steps. ∎

It is shown in [39] that some properties shared by a centrally symmetric $\mathcal{L}$ can be extended to those $\mathcal{L}$ whose corresponding matrices satisfy

$$B_{n,i}B_{n,j} = B_{n,j}B_{n,i}, \qquad 1 \le i,j \le d, \quad n \in \mathbb{N}_0. \tag{2.2.3}$$

We shall call linear functional whose corresponding matrices satisfy (2.2.3) *quasi-centrally symmetric*. Clearly, $B_{n,i} = 0$ implies (2.2.3); thus, central symmetry implies quasi-central symmetry. The following theorem provides further examples.

Theorem 2.2.2. *Let W be a weight function defined on $\Omega \subset \mathbf{R}^d$. Suppose that W becomes centrally symmetric under a nonsingular linear transformation*

$$\mathbf{u} \mapsto \mathbf{x}, \quad \mathbf{x} = T\mathbf{u} + \mathbf{a}, \quad \det T > 0, \quad \mathbf{x}, \mathbf{u} \in \mathbf{R}^d.$$

Then, the corresponding $B_{n,i}$ are in general nonzero and satisfy the condition (2.2.3).

Proof. Let $W^*(\mathbf{u}) = W(T\mathbf{x}+\mathbf{a})$ and $\Omega^* = \{\mathbf{u} : \mathbf{u} = T\mathbf{x}+\mathbf{a}, \quad \mathbf{x} \in \Omega\}$. If we denote by $\mathbf{P}_n$ the orthonormal polynomials associated to W, then following the change of variable $\mathbf{x} \mapsto T\mathbf{x} + \mathbf{a}$ it follows that the corresponding orthonormal polynomials for W^* are given by

$$\mathbf{P}_n^*(\mathbf{u}) = \sqrt{\det T}\, \mathbf{P}_n(T\mathbf{u} + \mathbf{a}).$$

Let $B_{n,i}(W)$ denote the matrices associated with W. Then, by (2.1.4)

$$\begin{aligned} B_{n,i}(W) &= \int_\Omega x_i \mathbf{P}_n(\mathbf{x})\mathbf{P}_n^T(\mathbf{x})W(\mathbf{x})d\mathbf{x} \\ &= \int_{\Omega^*} \sum_{j=1}^d (t_{ij}u_j + a_j)\mathbf{P}_n^*(\mathbf{u})\mathbf{P}_n^{*T}(\mathbf{u})W^*(\mathbf{u})d\mathbf{u} \\ &= \sum_{j=1}^d t_{ij}B_{n,j}(W^*) + a_i I. \end{aligned}$$

By assumption, W^* is centrally symmetric on Ω^*, which implies $B_{n,i}(W^*) = 0$ by Theorem 2.2.1. We conclude that $B_{n,i}(W) = a_i I$, which implies that $B_{n,i}(W)$ satisfies (2.2.3). ∎

In other words, linear functionals derived from weight functions which become centrally symmetric after a nonsingular affine transformation are quasi-centrally symmetric. This explains why we use the term "quasi". As examples, we mention that all product weight functions in Example 2.1 are quasi-centrally symmetric and they are centrally symmetric if μ' is even on a symmetric interval. However, it is not clear how to characterize quasi-central symmetry through the properties of the linear functional $\mathcal{L}$ or the weight function.

2.3. Lemmas

We need the definition of a generalized inverse of a matrix as stated in the first paragraph; several lemmas of general nature are given in the second part of this section.

2.3.1. Generalized inverse of a matrix. Let $A : s \times t$ be a matrix and assume $s \geq t$. We shall need the concept of a generalized inverse of a matrix in the special case where $\operatorname{rank} A = t$. Suppose that the singular value decomposition of A is given by

$$A = W^T \begin{pmatrix} \Sigma \\ \mathrm{O} \end{pmatrix} X$$

where $\Sigma : s \times s$ is an invertible diagonal matrix, $W : s \times s$ and $X : t \times t$ are orthogonal matrices, then a generalized inverse of A can be defined as

$$D^T = X^T(\Sigma^{-1}, \Sigma_1)W$$

where $\Sigma_1 : s - t \times t$ can be any matrix. It is easy to verify that D^T satisfies the equation

$$D^T A = I_t.$$

Because of the presence of Σ_1, the generalized inverse of A is not unique. When $\Sigma_1 = 0$, D^T is the unique Moore-Penrose generalized inverse; we denote it by A^+. In this paper, we reserve the letters U and V for a pair of matrices such that $U : s \times s - \sigma$ and $V : s \times \sigma$, $\sigma > 0$, satisfy

$$\operatorname{rank} U = s - \sigma, \quad \operatorname{rank} V = \sigma, \quad \text{and} \quad U^T V = 0.$$

It is easy to verify that the following matrix equation holds

$$\begin{pmatrix} U^T \\ V+ \end{pmatrix} ((U^+)^T, V) = I_s.$$

Since $((U^+)^T, V)$ is a square matrix, this shows that it is invertible. In particular, we can reverse the order of multiplication in the above equation which leads to

$$(U^+)^T U^T + VV^+ = I_s, \tag{2.3.1}$$

an equation that will be used repeatedly in Sections 4 and 7.

2.3.2. Lemmas. We let $L_{n,i}$ denote the matrices of size $r^d_{n-1} \times r^d_n$ which satisfy

$$L_{n,i}\mathbf{x}^n = x_i \mathbf{x}^{n-1}, \qquad 1 \le i \le d.$$

Clearly, $\operatorname{rank} L_{n,i} = r^d_{n-1}$, and $\operatorname{rank} L_n = r^d_n$, where $L_n = (L^T_{n,1}|\dots|L^T_{n,d})^T$. For example, for $d = 2$ we have

$$L_{n,1} = \begin{bmatrix} 1 & & \bigcirc & 0 \\ & \ddots & & \vdots \\ \bigcirc & & 1 & 0 \end{bmatrix} \quad \text{and} \quad L_{n,2} = \begin{bmatrix} 0 & 1 & & \bigcirc \\ \vdots & & \ddots & \\ 0 & \bigcirc & & 1 \end{bmatrix}.$$

By (2.1.4), the matrices $A_{n,i}$ and $L_{n,i}$ are related through the leading coefficient matrix G_n of $\mathbf{P}_n$ by

$$A_{n,i}G_{n+1} = G_n L_{n,i}, \qquad 1 \le i \le d. \tag{2.3.2}$$

For any given sequence of matrices $C_1, \dots, C_d$, where all C_i's are of the size $s \times t$, we define a matrix $\Xi_C : ds \times \binom{d}{2}t$ as follows. Let $\Xi_{i,j} : t \times ds$, $1 \le i, j \le d$, $i \ne j$, be block matrices defined by

$$\Xi_{i,j} = (\dots|0|C_j^T|\dots|-C_i^T|0|\dots); \tag{2.3.3}$$

i.e., the only two nonzero blocks are C_j^T at the ith block and $-C_i^T$at the jth block. The matrix Ξ_C is then defined by using $\Xi_{i,j}$ as blocks in the lexicographical order,

$$\Xi_C = [\Xi^T_{1,2}|\Xi^T_{1,3}|\dots|\Xi^T_{d-1,d}]. \tag{2.3.4}$$

For example, we have for $d = 4$,

$$\Xi_C = [\Xi^T_{1,2}|\Xi^T_{1,3}|\Xi^T_{1,4}|\Xi^T_{2,3}|\Xi^T_{2,4}|\Xi^T_{3,4}] = \begin{bmatrix} C_2 & C_3 & C_4 & 0 & 0 & 0 \\ -C_1 & 0 & 0 & C_3 & C_4 & 0 \\ 0 & -C_1 & 0 & -C_2 & 0 & C_4 \\ 0 & 0 & -C_1 & 0 & -C_2 & -C_3 \end{bmatrix}.$$

In particular, we have that the dimensions of $\Xi_{A_{n-1}}$ and $\Xi_{A^T_{n-1}}$ are, respectively,

$$\Xi_{A_{n-1}} : dr^d_{n-1} \times \binom{d}{2} r^d_n, \qquad \Xi_{A^T_{n-1}} : dr^d_n \times \binom{d}{2} r^d_{n-1}.$$

For example, for $d = 2$ and $d = 3$, respectively, we have that $\Xi_{A_{n-1}}$ is equal to

$$\begin{bmatrix} A_{n-1,2} \\ -A_{n-1,1} \end{bmatrix} \quad \text{and} \quad \begin{bmatrix} A_{n-1,2} & A_{n-1,3} & 0 \\ -A_{n-1,1} & 0 & A_{n-1,3} \\ 0 & -A_{n-1,1} & -A_{n-1,2} \end{bmatrix}.$$

In [39] we computed the ranks of Ξ_{L_n} and $\Xi_{L^T_n}$, the results are as follows.

Lemma 2.3.1. *For $d \geq 2$ and $n \geq 1$,*

$$\operatorname{rank} \Xi_{L_n} = dr^d_{n-1} - r^d_{n-2}, \quad \text{and} \quad \operatorname{rank} \Xi_{L^T_n} = dr^d_n - r^d_{n+1}.$$

Using (2.3.2) we can find the rank of the corresponding matrices for $A_{n-1,i}$.

Lemma 2.3.2. *For $d \geq 2$ and $n \geq 1$,*

$$\operatorname{rank} \Xi_{A_{n-1}} = dr^d_{n-1} - r^d_{n-2}, \quad \text{and} \quad \operatorname{rank} \Xi_{A^T_{n-1}} = dr^d_n - r^d_{n+1}.$$

Proof. By (2.3.2) and the definition of Ξ_C it readily follows that

$$\Xi_{A_{n-1}} \operatorname{diag}\{G_n, \ldots, G_n\} = \operatorname{diag}\{G_{n-1}, \ldots, G_{n-1}\} \Xi_{L_n},$$

where the block diagonal matrix on the left-hand side is of size $\binom{d}{2} r^d_n \times \binom{d}{2} r^d_n$, and the one on the right-hand side is of size $dr^d_{n-1} \times dr^d_{n-1}$. Since G_n and G_{n-1} are both invertible, so are these two block ones. Therefore, we have $\operatorname{rank} \Xi_{A_{n-1}} = \operatorname{rank} \Xi_{L_n}$. Thus, the first equality follows from the corresponding one in Lemma 2.3.1. The second one is proved similarly. ∎

Our next lemma plays an important role in our investigation. It was proved in [39] for $D^T_n = A^+_n$, but since we need to use the lemma in more general situations, a different proof is given.

Lemma 2.3.3. *Let $A_{n,i}$ be the coefficient matrices in the three-term relation (2.1.2), and let D^T_n satisfies (2.1.8). Then $Y \in \mathbb{R}^{dr^d_n}$ is in the null space of $I - A_n D^T_n$ if, and only if, $\Xi^T_{A^T_{n-1}} Y = 0$. Moreover, if $Y = (Y^T_1, \ldots, Y^T_d)$, $Y_i \in \mathbb{R}^{r^d_n}$, then $\Xi^T_{A^T_{n-1}} Y = 0$ is equivalent to*

$$A_{n-1,i} Y_j = A_{n-1,j} Y_i, \quad 1 \leq i < j \leq d. \tag{2.3.5}$$

Proof. Let the singular value decomposition of A_n and D^T_n be of the form given in **2.3.1**. Both A_n and D_n are of size $dr^d_n \times r^d_{n+1}$. It follows readily that

$$I - A_n D^T_n = W^T \begin{pmatrix} \mathbb{O} & -\Sigma\Sigma_1 \\ \mathbb{O} & I \end{pmatrix} W,$$

where I is the identity matrix of size $dr_n^d - r_{n+1}^d$, from which it follows that

$$\operatorname{rank}(I - A_n D_n^T) = dr_n^d - r_{n+1}^d.$$

Hence,

$$\dim \ker(I - A_n D_n^T) = dr_n^d - \operatorname{rank}(I - A_n D_n^T) = r_{n+1}^d.$$

By (2.1.8) it follows that $(I - A_n D_n^T)A_n = A_n - A_n = 0$, which implies that the columns of A_n form a basis for the null space of $I - A_n D_n^T$. Therefore, if $(I - A_n D_n^T)Y = 0$, then there exists a vector $\mathbf{c}$ such that $Y = A_n\mathbf{c}$. But by the commuting conditions (2.1.5),

$$\Xi_{A_{n-1}^T}^T A_n = 0; \tag{2.3.6}$$

it follows that $\Xi_{A_{n-1}^T}^T Y = 0$. On the other hand, suppose $\Xi_{A_{n-1}^T}^T Y = 0$. By Lemma 2.3.2,

$$\dim \ker \Xi_{A_{n-1}^T}^T = dr_n^d - \operatorname{rank} \Xi_{A_{n-1}^T}^T = r_{n+1}^d.$$

From (2.3.6) it also follows that the columns of A_n form a basis for $\ker \Xi_{A_{n-1}^T}^T$. Therefore, there exists a vector $\mathbf{c}^*$ such that $Y = A_n\mathbf{c}^*$, which, by (2.1.8), implies that $(I - A_n D_n^T)Y = 0$. The equivalence of $\Xi_{A_{n-1}^T}^T Y = 0$ and the equations in (2.3.5) follows readily from the definition of $\Xi_{A_{n-1}}$ in (2.3.4). ∎

3. Motivations

In this section we present two observations that motivated our study; both of them are of independent interests. The first one is concerned with the zeros of $\mathbb{P}_n$ for certain special linear functionals including the centrally symmetric ones. The second one is about necessary conditions of a minimal cubature formula which motivates the classes of polynomials whose zeros we shall study. These polynomial classes are formally defined in the last subsection.

3.1. Zeros for a special functional

By Mysovskikh's theorem Gaussian cubature exists if, and only if, $\mathbb{P}_n$ has $N = \dim \Pi^d_{n-1}$ many zeros. In general, $\mathbb{P}_n$ can have at most these many zeros. For a minimal cubature formula, it seems natural to expect the common zeros of $\mathbb{P}_n$, though surely not enough in number, to play a substantial role. By Theorem 1.1, it is readily seen that $\mathbb{P}_n$ can have at most $\dim \Pi^d_{n-1} - r_{i,j}$ many zeros in general, where

$$r_{i,j} = \text{rank}(A_{n,i}A^T_{n,j} - A_{n,j}A^T_{n,i}).$$

We note that this matrix, being skew symmetric, must have even rank. Nevertheless, we have the following theorem which seems to be surprising.

Theorem 3.1.1. *Let $d = 2$ and suppose that*

$$\text{(3.1.1)} \qquad \text{rank}(A_{n-1,1}A^T_{n-1,2} - A_{n-1,2}A^T_{n-1,1}) = 2\left[\frac{r^2_{n-1}}{2}\right] = 2\left[\frac{n}{2}\right];$$

i.e., the matrix has full rank. Then for even integer n, $\mathbb{P}_n$ has no zeros. Moreover, if $\mathcal{L}$ is centrally symmetric, then for odd integer n, $\mathbb{P}_n$ has $\Lambda = 0$ as the only zero.

Proof. By Theorem 1.1, if Λ is a zero of $\mathbb{P}_n$, then it is a joint eigenvalue of $T_{n,i}$ with common eigenvector $\mathbf{x}_\Lambda = (\mathbb{P}^T_0(\Lambda), \mathbb{P}^T_1(\Lambda), \ldots, \mathbb{P}^T_{n-1}(\Lambda))^T$. In particular,

$$(T_{n,1}T_{n,2} - T_{n,2}T_{n,1})\mathbf{x}_\Lambda = (\lambda_1\lambda_2 - \lambda_2\lambda_1)\mathbf{x}_\Lambda = 0.$$

By the commuting conditions (2.1.5), (2.1.6), and (2.1.7), it follows readily that this equation is equivalent to

$$(A_{n-1,1}A_{n-1,2}^T - A_{n-1,2}A_{n-1,1}^T)\mathbf{P}_{n-1}(\Lambda) = 0.$$

If n is even, then (3.1.1) implies that the matrix is invertible, thus, we have $\mathbf{P}_{n-1}(\Lambda) = 0$. However, since it is known that $\mathbf{P}_n$ and $\mathbf{P}_{n-1}$ do not have common zeros, this leads to a contradiction, which shows that $\mathbf{P}_n$ does not have any zero.

On the other hand, if n is odd, then (3.1.1) implies that the matrix is of dimension $n-1$ and its null space is spanned by one nonzero vector. If, in addition, $\mathcal{L}$ is centrally symmetric, then this vector can be taken to be $\mathbf{P}_{n-1}(0)$. Indeed, for centrally symmetric $\mathcal{L}$, we have $B_{n,i} = 0$, thus, for $\Lambda = 0$ the three-term relation becomes

$$A_{k-1,i}^T\mathbf{P}_{k-1}(0) + A_{k,i}\mathbf{P}_{k+1}(0) = 0, \qquad k \geq 1, \tag{3.1.2}$$

and $A_{0,i}\mathbf{P}_1(0) = 0$. Therefore, for odd n, using (2.1.8) it follows from these equations recursively that $\mathbf{P}_n(0) = 0$. Moreover, by (3.1.2) and the commuting condition (2.1.5), we have in this case

$$(A_{n-1,1}A_{n-1,2}^T - A_{n-1,2}A_{n-1,1}^T)\mathbf{P}_{n-1}(0) = -(A_{n-1,1}A_{n,2} - A_{n-1,2}A_{n,1})\mathbf{P}_{n+1}(0) = 0.$$

Since $\mathbf{P}_n(0) = 0$, we know $\mathbf{P}_{n-1}(0) \neq 0$. Therefore, $\mathbf{P}_{n-1}(0)$ spans the null space of the matrix in (3.1.1). ∎

The condition (3.1.1) is known to be satisfied for quasi-centrally symmetric linear functionals ([18, 39], see Section 5). Part of the proof of the above theorem is independent of dimension. We have the following result which seems to be of some interest.

Theorem 3.1.2. *Let $\mathcal{L}$ be a centrally symmetric linear functional defined on Π^d. If n is odd, then 0 is a zero of $\mathbf{P}_n$.*

Clearly, the proof follows easily from (2.1.8) and (3.1.2). This theorem will be used later in Section 5.

3.2. Necessary conditions for the existence of minimal cubature formula

Our second observation is concerned with minimal cubature formulae of degree $2n-1$. By (1.2.9) such a formula will have $N \geq \dim \Pi_{n-1}^d$ many nodes, which follows quite easily from the fact that no polynomial of degree $n-1$ can vanish on all nodes without violating the square positivity of the linear functional. Since we are concerned with formulae of degree $2n-1$, we write

$$(3.2.1) \qquad N = \dim \Pi_{n-1}^d + \sigma, \qquad 0 \leq \sigma < r_n^d, \quad \sigma \in \mathbb{N}.$$

The restriction that $\sigma < r_n^d$ follows from the fact that

$$\dim \Pi_n^d = \dim \Pi_{n-1}^d + r_n^d,$$

and from the fact that $N \geq \dim \Pi_n^d$ many nodes may yield a cubature formula of degree $2n+1$ instead of $2n-1$. For a minimal cubature formula, we write $\sigma = \sigma_{\min}$.

Let $\mathbf{x}_1, \ldots, \mathbf{x}_N$ be nodes of a cubature formula of degree $2n-1$, see (1.2.7). For $m \in \mathbb{N}_0$ we define a matrix

$$(3.2.2) \qquad \Psi_m = (\mathbf{x}_k^\alpha)_{|\alpha| \leq m}{}_{k=1}^{N},$$

where the rows of Ψ_m are numbered according to the lexicographical order in the set $\{\alpha \in \mathbb{N}^d, |\alpha| \leq m\}$ and the columns are numbered according to the nodes $\mathbf{x}_1, \ldots, \mathbf{x}_N$. We start with a simple lemma:

Lemma 3.2.1. *Let $\mathbf{x}_1, \ldots, \mathbf{x}_N$ be the nodes of a minimal cubature formula of degree $2n-1$. Then the matrix Ψ_{2n-1} has full rank; i.e.,*

$$(3.2.3) \qquad \operatorname{rank} \Psi_{2n-1} = N.$$

Proof. Suppose otherwise, then there exists a vector $\mathbf{a} \in \mathbb{R}^N$ such that $\Psi \mathbf{a} = 0$, which implies that

$$\sum_{k=1}^{N} a_k \mathbf{x}_k^\alpha = 0, \qquad 0 \leq |\alpha| \leq 2n-1.$$

Therefore, summing over α we have

$$\sum_{k=1}^{N} P(\mathbf{x}_k) = 0, \qquad \forall P \in \Pi_{2n-1}^d.$$

Let $\mu = \min\{\Lambda_i a_i^{-1} : a_i > 0\}$, where Λ_i's are the weights of the minimal cubature formula. Then we have

$$\sum_{k=1}^{N} (\Lambda_k - \mu a_k) P(\mathbf{x}_k) = \mathcal{L}(P), \qquad P \in \Pi_{2n-1}^d.$$

However, since $\Lambda_k - \mu a_k \geq 0$ and at least one of them is zero, this is a cubature formula with $N-1$ nodes, which contradicts the fact that $\mathbf{x}_1 \ldots, \mathbf{x}_N$ are the nodes of a minimal cubature formula. ∎

This property of matrix Ψ_{2n-1} is equivalent to what is known as the "interpolatory property" of a minimal cubature formula. Though this lemma is known, we gave this short proof for the sake of completeness.

We note that $\operatorname{rank} \Psi_{n-1} = \dim \Pi_{n-1}^d$ as there is no polynomial in Π_{n-1}^d vanishing on all nodes. When Gaussian cubature of degree $2n-1$ exists, we have $N = \dim \Pi_{n-1}^d$ and $\operatorname{rank} \Psi_n = N$. In general, though, the rank of Ψ_m is increasing from $\operatorname{rank} \Psi_{n-1} = \dim \Pi_{n-1}^d$ to $\operatorname{rank} \Psi_{2n-1} = \dim \Pi_{n-1}^d + \sigma_{\min}$. We define a sequence of numbers $\sigma_0, \ldots, \sigma_{n-1}$ as follows.

(3.2.4) $$\operatorname{rank} \Psi_{n+k} = \dim \Pi_{n-1}^d + \sigma_0 + \ldots + \sigma_k, \qquad 0 \leq k \leq n-1;$$

the reason that we use $\sigma_0 + \ldots + \sigma_k$ instead of a single number will become clear in the next theorem. From this definition, we clearly have $\sigma_k \geq 0$. By Lemma 3.2.1, we have

$$\sigma_{\min} = \sigma_0 + \ldots + \sigma_{2n-1}.$$

Theorem 3.2.2. *Let $\mathbf{x}_1, \ldots, \mathbf{x}_N$ be the nodes of a minimal cubature formula of degree $2n-1$ and assume (3.2.4). Then for $0 \leq k \leq n-1$, there are precisely $r_{n+k}^d - \sigma_k$ many linear independent polynomials of degree exactly $n+k$ which vanish on all $\mathbf{x}_j$, $1 \leq j \leq N$. Moreover, these polynomials are orthogonal to Π_{n-k-1}^d.*

Proof. Let us consider the matrix Ψ_n first. By (3.2.4) we have

$$\dim \ker \Psi_n^T = \dim \Pi_n^d - (\dim \Pi_{n-1}^d + \sigma_0) = r_n^d - \sigma_0.$$

Therefore, there exist $r_n^d - \sigma_0$ many linearly independent vectors $\mathbf{b}_i$, $1 \le i \le r_n^d - \sigma_0$, such that $\Psi_n^T \mathbf{b}_i = 0$. These vectors induce a set of $r_n^d - \sigma_0$ many linearly independent polynomials $Q_i \in \Pi_n^d$, defined by $Q_i = (1, \mathbf{x}^T, \ldots, (\mathbf{x}^n)^T)\mathbf{b}_i$, $1 \le i \le r_n^d - \sigma_0$, which vanish on all $\mathbf{x}_k$, $1 \le k \le N$. Since the cubature formula is of degree $2n-1$, we have for any polynomial $P \in \Pi_{n-1}^d$

$$\mathcal{L}(Q_i P) = \sum_{k=1}^{N} Q_i(\mathbf{x}_k) P(\mathbf{x}_k) = 0,$$

which shows that the Q_i's are orthogonal to Π_{n-1}^d. In general, for $k \ge 0$ we have by (3.2.4)

$$\begin{aligned}\dim\ker \Psi_{n+k}^T &= \dim \Pi_{n+k}^d - (\dim \Pi_{n-1}^d + \sigma_0 + \ldots + \sigma_k) \\ &= r_{n+k}^d + \ldots + r_n^d - (\sigma_0 + \ldots + \sigma_k).\end{aligned}$$

Following the argument given above, we can conclude that there are exactly $r_{n+k}^d + \ldots + r_n^d - (\sigma_0 + \ldots + \sigma_k)$ many linearly independent polynomials in Π_{n+k}^d which vanish on all nodes of the cubature formula. Among these polynomials, there are $r_{n+k-1}^d + \ldots + r_n^d - (\sigma_0 + \ldots + \sigma_{k-1})$ many which, corresponding to a basis in $\ker \Psi_{n+k-1}^T$, are polynomials in Π_{n+k-1}^d. Therefore, there are exactly $r_{n+k}^d - \sigma_k$ many linearly independent polynomials in Π_{n+k}^d which vanish on all nodes of the formula. Moreover, since the formula is exact for Π_{2n-1}^d, these polynomials are orthogonal to polynomials of degree $(2n-1) - (n+k) = n-k-1$. This completes the proof. ∎

The theorem indicates that the nodes of a minimal cubature formula are common zeros of a family of quasi-orthogonal polynomials of consecutive orders. Using the notation introduced in **2.1.3**, we denote the quasi-orthogonal polynomials of degree $n+k$ which are orthogonal to Π_{n-1+k}^d as

$$\mathbf{Q}_{n+k} = \mathbf{P}_{n+k} + \Gamma_{1,k}\mathbf{P}_{n+k-1} + \ldots + \Gamma_{2k,k}\mathbf{P}_{n-k}. \tag{3.2.5}$$

Here and in the following, we take n fixed and write $\mathbf{Q}_{n+k}$ instead of $\mathbf{Q}_{n+k,2k}$. Using this vector notation, we reformulate Theorem 3.2.2 as follows:

Theorem 3.2.3. *Let $\mathbf{x}_1, \ldots, \mathbf{x}_N$ be the nodes of a minimal cubature formula of degree $2n-1$ and assume (3.2.4). Then for $0 \le k \le n-1$, there exist matrices $U_k : r^d_{n+k} \times r^d_{n+k} - \sigma_k$ of full rank, such that $\mathbf{x}_1, \ldots, \mathbf{x}_N$ are common zeros of $U_0^T \mathbf{P}_n, U_1^T \mathbf{Q}_{n+1}, \ldots, U_{n-1}^T \mathbf{Q}_{2n-1}$.*

If a Gaussian cubature formula exists, then we have $N = \dim \Pi^d_{n-1}$ and $\sigma_0 = \ldots = \sigma_{n-1} = 0$; the nodes are common zeros of $\mathbf{P}_n$. In general, it is natural to expect that some of the σ_k's will be zero. From the proof of Theorem 3.2.2, if $\sigma_r = 0$ for some r then $\operatorname{rank} \Psi_{n+r} = N$, which implies immediately that $\sigma_{r+1} = \ldots = \sigma_{n-1} = 0$. In particular, the simplest case $\sigma_1 = 0$ means that we only need to study the common zeros of $U_0^T \mathbf{P}_n$ and $\mathbf{Q}_{n+1}$.

3.3. Definitions

The main task we will undertake in this paper is to study the common zeros of quasi-orthogonal polynomials. More precisely, we will study the common zeros of the polynomials in the set

$$\mathcal{Q} = \{U_0^T \mathbf{P}_n, U_1^T \mathbf{Q}_{n+1}, \ldots, U_r^T \mathbf{Q}_{n+r}, \mathbf{Q}_{n+r+1}\}, \tag{3.3.1}$$

where the $\mathbf{Q}_k$'s are defined in (3.2.5) and the U_k, $1 \le k \le r$, are matrices such that

$$U_k : r^d_{n+k} \times r^d_{n+k} - \sigma_k, \qquad \operatorname{rank} U = r^d_{n+k} - \sigma_k. \tag{3.3.2}$$

As before, we call the common zeros simply zeros of $\mathcal{Q}$. For convenience we also define $\mathbf{Q}_n = \mathbf{P}_n$ and $U_{r+1} = I$ to write $\mathcal{Q}$ more compactly as

$$\mathcal{Q} = \{U_0^T \mathbf{Q}_n, \ldots, U_{r+1}^T \mathbf{Q}_{n+r+1}\}, \qquad r \ge 0. \tag{3.3.3}$$

For a minimal cubature formula, as shown in the previous section, if $\mathbf{Q}_{n+r}$ vanishes on all nodes, then there exist $\mathbf{Q}_{n+r+k}$, $k \ge 1$, such that all $\mathbf{Q}_{n+r+k}$ vanish on these nodes. In particular, the zero set of $\mathcal{Q}$ in (3.3.3) is identical to that of $\mathcal{Q}$ with $\mathbf{Q}_{n+r+k}$ added. Although this will not cause problems in our study of zeros of $\mathcal{Q}$, as we shall see in Section 7, it may cause confusion in our discussion of cubature. Therefore, we make the following definitions.

Definition 3.3.1. *If a cubature formula is based on the zeros of $\mathcal{Q}$, where $\mathcal{Q}$ is defined as in (3.3.1) and U_r is not the identity matrix, then we say that the cubature formula is generated by the zeros of $\mathcal{Q}$.*

There is another definition which turns out to be quite useful:

Definition 3.3.2. *Let $\mathcal{Q}$ be defined as above. We call $\mathcal{Q}$ maximal, if for $1 \leq k \leq r$ every polynomial in Π_{n+k}^d vanishing on all the zeros of $\mathcal{Q}$ belongs to the linear space spanned by $U_0^T \mathbb{Q}_n, \dots, U_{n+k}^T \mathbb{Q}_{n+k}$.*

Clearly, for a minimal cubature formula, $\mathcal{Q}$ in Theorem 3.2.3 is maximal. The meaning of this notion will become clear in our next section.

4. Common Zeros of Polynomials in Several Variables: First Case

In this section we study the first case in view of the definitions given in Section 3.3: common zeros of $U_0^T \mathbf{P}_n$ and $\mathbf{Q}_{n+1}$, and some of its consequences. To simplify notation we drop the subindex to write U for U_0, recall that

$$U : r_n^d \times r_n^d - \sigma, \qquad \operatorname{rank} U = r_n^d - \sigma. \tag{4.1}$$

We let $V : r_n^d \times \sigma$ be the matrix associated with U as defined in **2.3.1**,

$$U^T V = 0, \qquad \text{and} \qquad \operatorname{rank} V = \sigma. \tag{4.2}$$

We also write

$$\mathbf{Q}_{n+1} = \mathbf{P}_{n+1} + \Gamma_1 \mathbf{P}_n + \Gamma_2 \mathbf{P}_{n-1}. \tag{4.3}$$

One of the major difficulties we face is that $U^T \mathbf{P}_n$ only spans a subspace of the space spanned by $\mathbf{P}_n$, yet the three-term relation and all what follows from it are more or less based on the whole space spanned by $\mathbf{P}_n$. In proving Theorem 1.1 and related results in [38] we looked at the corresponding result in one variable for guidance; if we would do so in the present more general setting, the difficulties would be unprecedented. In following the approach in one variable, where the space spanned by $\mathbf{P}_n$ has dimension one, we take $\mathbf{P}_n$ as a whole, but now we have to divide it into parts in order to understand the situation. Nevertheless, it turns out that we can follow the approach we used in proving Theorem 1.1 as a general guideline. First we show that the common zeros of $U^T \mathbf{P}_n$ and $\mathbf{Q}_{n+1}$ are eigenvalues of truncated block Jacobi matrices which are properly tailored. Next, we show that the matrices can be made symmetric, so that the zeros are real, and commuting, which implies that there is a maximal set of zeros. The central theorem, Theorem 4.1.4, gives a complete characterization for $U^T\mathbf{P}_n$ and $\mathbf{Q}_{n+1}$ to have the maximal number of distinct real common zeros. That the zeros are distinct is proved through a modified Christoffel-Darboux formula, which is also used in Lagrange interpolation and in cubature. Assuming that zeros exist, we prove uniqueness of a special Lagrange interpolation problem based on these zeros which, upon integration, leads to a cubature formula of degree $2n-1$.

4.1. Characterization of zeros

Using the matrix V we define a family of truncated block Jacobi matrices associated with U as follows.

$$S_{n,i} = \begin{bmatrix} B_{0,i} & A_{0,i} & & & \bigcirc \\ A_{0,i}^T & B_{1,i} & A_{1,i} & & \\ & \ddots & \ddots & \ddots & \\ & & & B_{n-2,i} & A_{n-1,i}V \\ \bigcirc & & & V^+A_{n-1,i}^{*T} & V^+B_{n,i}^*V \end{bmatrix}, \qquad 1 \le i \le d, \tag{4.1.1}$$

where the matrices $A_{n-1,i}^*$ and $B_{n,i}^*$ are defined by

$$A_{n-1,i}^{*T} = A_{n-1,i}^T - A_{n,i}\Gamma_2, \tag{4.1.2}$$

$$B_{n,i}^* = B_{n,i} - A_{n,i}\Gamma_1. \tag{4.1.3}$$

We say that $\Lambda = (\lambda_1, \ldots \lambda_d)$ is an eigenvalue of the family of matrices $S_{n,1}, \ldots, S_{n,d}$ if, and only if, there is a nonzero vector $\mathbf{x}$ such that

$$S_{n,i}\mathbf{x} = \lambda_i \mathbf{x}, \qquad 1 \le i \le d.$$

This vector $\mathbf{x}$ is called a joint eigenvector of Λ. We begin with

Lemma 4.1.1. *If $\Lambda = (\lambda_1, \ldots, \lambda_d)$ is a common zero of $U^T\mathbf{P}_n$ and $\mathbf{Q}_{n+1}$, then Λ is a joint eigenvalue of $S_{n,1}, \ldots, S_{n,d}$ with a joint eigenvector given by*

$$[\mathbf{P}_0^T(\Lambda), \ldots, \mathbf{P}_{n-1}^T(\Lambda), (V^+\mathbf{P}_n(\Lambda))^T]^T.$$

Proof. Suppose Λ is a common zero of $U^T\mathbf{P}_n$ and $\mathbf{Q}_{n+1}$. By (2.3.1) and $U^T\mathbf{P}_n(\Lambda) = 0$, it readily follows that

$$\mathbf{P}_n(\Lambda) = ((U^+)^T U^T + VV^+)\mathbf{P}_n(\Lambda) = VV^+\mathbf{P}_n(\Lambda).$$

By the three-term relation (2.1.2) it further follows that

$$\begin{aligned} B_{0,i}\mathbf{P}_0(\Lambda) + A_{0,i}\mathbf{P}_1(\Lambda) &= \lambda_i\mathbf{P}_0(\Lambda) \\ A_{k-1,i}^T\mathbf{P}_{k-1}(\Lambda) + B_{k,i}\mathbf{P}_k(\Lambda) + A_{k,i}\mathbf{P}_{k+1}(\Lambda) &= \lambda_i\mathbf{P}_k(\Lambda), \quad 1 \le k \le n-2 \\ A_{n-2,i}^T\mathbf{P}_{n-2}(\Lambda) + B_{n-1,i}\mathbf{P}_{n-1}(\Lambda) + A_{n-1,i}VV^+\mathbf{P}_n(\Lambda) &= \lambda_i\mathbf{P}_{n-1}(\Lambda) \end{aligned}$$

for $1 \le i \le d$. Since $\mathbf{Q}_{n+1}(\Lambda) = 0$ implies

$$\mathbf{P}_{n+1}(\Lambda) = -\Gamma_1\mathbf{P}_n(\Lambda) - \Gamma_2\mathbf{P}_{n-1}(\Lambda),$$

we have from (2.1.2)

$$(A_{n-1,i}^T - A_{n,i}\Gamma_1)\mathbf{P}_{n-1}(\Lambda) + (B_{n,i} - A_{n,i}\Gamma_2)VV^+\mathbf{P}_n(\Lambda) = \lambda_i\mathbf{P}_n(\Lambda).$$

Multiply the last equation by V^+ and use the definition of $S_{n,i}$, we see that these equations are equivalent to

$$S_{n,i}\mathbf{x} = \lambda_i\mathbf{x}, \quad \mathbf{x} = [\mathbf{P}_0^T(\Lambda), \ldots, \mathbf{P}_{n-1}^T(\Lambda), (V^+\mathbf{P}_n(\Lambda))^T]^T, \qquad 1 \le i \le d.$$

Thus, Λ is an eigenvalue of $S_{n,1}, \ldots, S_{n,d}$ with $\mathbf{x}$ as joint eigenvector. ∎

One interesting consequence of the lemma is the following theorem.

Theorem 4.1.2. *For any U and $\mathbf{Q}_{n+1}$ defined in (4.1) and (4.3), the set $Q = \{U^T\mathbf{P}_n, \mathbf{Q}_{n+1}\}$ has at most $N := \dim \Pi_{n-1}^d + \sigma$ distinct zeros.*

Proof. Since $S_{n,i}$ is a square matrix of the size $\dim \Pi_{n-1}^d + \sigma$, there are at most $\dim \Pi_{n-1}^d + \sigma$ many joint eigenvectors of $S_{n,1}, \ldots S_{n,d}$. Therefore the theorem follows readily from Lemma 4.1.1. ∎

In the theorem there is no restriction put on the matrices Γ_1 and Γ_2, and very little on V. Our next theorem, which is the central result in this section, provides necessary and sufficient conditions posed on these matrices so that Q has the maximal number of zeros. We need a lemma first.

Lemma 4.1.3. *If $\mathcal{Q} = \{U^T\mathbf{P}_n, \mathbf{Q}_{n+1}\}$ is maximal, then*

$$U^T A_{n-1}^{*T} = 0, \quad \text{and} \quad U^T B_{n,i}^* V = 0, \qquad 1 \le i \le d. \tag{4.1.4}$$

Proof. The fact that $\mathcal{Q}$ is maximal implies that any polynomial of degree $n+1$ which vanishes on the zeros of $\mathcal{Q}$ must belong to the linear space spanned by $U^T\mathbf{P}_n$ and $\mathbf{Q}_{n+1}$. This applies, in particular, to the polynomials $x_i U^T\mathbf{P}_n$; we have

$$x_i U^T \mathbf{P}_n = H_{1,i}\mathbf{Q}_{n+1} + H_{2,i} U^T \mathbf{P}_n, \quad 1 \le i \le d,$$

for some matrices $H_{1,i}$ and $H_{2,i}$. Multiplying this equation by $\mathbf{P}_{n+1}^T$, $\mathbf{P}_n^T$, and $\mathbf{P}_{n-1}^T$, respectively, and applying the linear functional $\mathcal{L}$, we obtain by (2.1.4)

$$H_{1,i} = U^T A_{n,i}, \quad U^T B_{n,i}^* = H_{2,i} U^T, \quad \text{and} \quad U^T A_{n-1,i}^T = U^T A_{n,i}\Gamma_2.$$

By (4.1.2) the last equation is equivalent to the first equation in (4.1.4), and by $U^T V = 0$ the second equation implies the second one in (4.1.4). ∎

Theorem 4.1.4. *The set $\mathcal{Q} = \{U^T\mathbf{P}_n, \mathbf{Q}_{n+1}\}$ has $\dim \Pi_{n-1}^d + \sigma$ many pairwise distinct real zeros and $\mathcal{Q}$ is maximal if, and only if,*

$$\Gamma_2 = \sum_{i=1}^{d} D_{n,i}^T (I - VV^T) A_{n-1,i}^T, \tag{4.1.5}$$

and Γ_1 and V satisfy the following conditions:

$$A_{n-1,i}(VV^T - I)A_{n-1,j}^T = A_{n-1,j}(VV^T - I)A_{n-1,i}^T, \qquad 1 \le i,j \le d, \tag{4.1.6}$$

$$(B_{n,i} - A_{n,i}\Gamma_1)VV^T = VV^T(B_{n,i} - \Gamma_1^T A_{n,i}^T), \qquad 1 \le i \le d, \tag{4.1.7}$$

$$\begin{aligned} VV^T A_{n-1,i}^T A_{n-1,j} VV^T + &(B_{n,i} - A_{n,i}\Gamma_1)VV^T(B_{n,j} - \Gamma_1^T A_{n,j}^T) = \\ VV^T A_{n-1,j}^T A_{n-1,i} VV^T + &(B_{n,j} - A_{n,j}\Gamma_1)VV^T(B_{n,i} - \Gamma_1^T A_{n,i}^T), \quad 1 \le i,j \le d. \end{aligned} \tag{4.1.8}$$

Proof. First, suppose that $U^T\mathbf{P}_n$ and $\mathbf{Q}_{n+1}$ have $\dim \Pi_{n-1}^d + \sigma$ many real distinct common zeros. Then, by Lemma 4.1.1, the matrices $S_{n,1}, \ldots, S_{n,d}$ have these many distinct joint eigenvalues. The matrices have to be symmetric and simultaneously diagonalizable. The symmetry of $S_{n,i}$ implies

$$V^+ A^{*T}_{n-1,i} = V^T A^T_{n-1,i}, \qquad 1 \le i \le d, \tag{4.1.9}$$

and

$$V^+ B^*_{n,i} V = V^T B^{*T}_{n,i} (V^+)^T, \qquad 1 \le i \le d. \tag{4.1.10}$$

Since Q is maximal, we obtain from (4.1.4) and (2.3.1) that (4.1.9) implies

$$A^{*T}_{n-1,i} = VV^T A^T_{n-1,i},$$

which, by the definition of $A^*_{n-1,i}$ in (4.1.2), is equivalent to

$$A_{n,i}\Gamma_2 = (I - VV^T) A^T_{n-1,i}, \qquad 1 \le i \le d.$$

Applying (2.1.8), we derive immediately from this equation the formula given for Γ_2 in (4.1.5). Again by (4.1.4) and (2.3.1), we have that (4.1.10) implies

$$B^*_{n,i} V = VV^T B^{*T}_{n,i} (V^+)^T,$$

which, upon multiplying by V^T and using (4.1.4) and (2.3.1) again, is equivalent to (4.1.7). Taking (4.1.9) and (4.1.10) into consideration and using the commuting conditions (2.1.5), (2.1.6), and (2.1.7), we have that

$$S_{n,i} S_{n,j} = S_{n,j} S_{n,i}$$

are equivalent to the equations

$$\begin{aligned} A^T_{n-2,i} A_{n-2,j} + B_{n-1,i} B_{n-1,j} + A_{n-1,i} VV^T A^T_{n-1,j} \\ = A^T_{n-2,j} A_{n-2,i} + B_{n-1,j} B_{n-1,i} + A_{n-1,j} VV^T A^T_{n-1,i}, \end{aligned} \tag{4.1.11}$$

$$\begin{aligned} B_{n-1,i} A_{n-1,j} V + A_{n-1,i} VV^+ B^*_{n,j} V \\ = B_{n-1,j} A_{n-1,i} V + A_{n-1,j} VV^+ B^*_{n,i} V, \end{aligned} \tag{4.1.12}$$

$$(4.1.13)\qquad V^TA^T_{n-1,i}A_{n-1,j}V + V^+B^*_{n,i}VV^+B^*_{n,j}V = V^TA^T_{n-1,j}A_{n-1,i}V + V^+B^*_{n,j}VV^+B^*_{n,i}V.$$

By (2.1.7) it follows that (4.1.11) is equivalent to (4.1.6). By (2.3.1) we get from (4.1.4) that

$$(4.1.14)\qquad VV^+B^*_{n,i}V = B^*_{n,i}V, \qquad 1 \le i \le d,$$

which, together with (4.1.7), implies that (4.1.13) is equivalent to (4.1.8). Thus, the necessity of all conditions have been verified.

On the other hand, assume that Γ_2 is given as in (4.1.5) and conditions (4.1.6), (4.1.7), and (4.1.8) are satisfied. First, we prove that every eigenvalue of $S_{n,1},\ldots,S_{n,d}$ which has a joint eigenvector $\mathbf{x}$ must be a zero of both $U^T\mathbf{P}_n$ and $\mathbf{Q}_{n+1}$. Suppose that $\Lambda = (\lambda_1,\ldots,\lambda_d)$ is such an eigenvalue and its eigenvector is written as

$$\mathbf{x} = (\mathbf{x}_0^T,\ldots,\mathbf{x}_{n-1}^T,\mathbf{x}_n^T)^T, \quad \mathbf{x}_j \in \mathbb{R}^{r_j^d}, \quad 1 \le j \le n-1, \quad \text{and} \quad \mathbf{x}_n \in \mathbb{R}^{\sigma}.$$

Since $S_{n,i}\mathbf{x} = \lambda_i\mathbf{x}$, it follows that $\{\mathbf{x}_j\}$ satisfies a three-term relation

$$(4.1.15)\qquad \begin{aligned} B_{0,i}\mathbf{x}_0 + A_{0,i}\mathbf{x}_1 &= \lambda_i\mathbf{x}_0 \\ A^T_{k-1,i}\mathbf{x}_{k-1} + B_{k,i}\mathbf{x}_k + A_{k,i}\mathbf{x}_{k+1} &= \lambda_i\mathbf{x}_k, \quad 1 \le k \le n-2 \\ A^T_{n-2,i}\mathbf{x}_{n-2} + B_{n-1,i}\mathbf{x}_{n-1} + A_{n-1,i}Vx_n &= \lambda_i\mathbf{x}_{n-1} \\ V^+A^{*T}_{n-1,i}\mathbf{x}_{n-1} + V^+B^*_{n,i}\mathbf{x}_n &= \lambda_i\mathbf{x}_n \end{aligned}$$

for $1 \le i \le d$. We show that $\mathbf{x}_0$, thus $\mathbf{x}$, is not zero. Indeed, if $\mathbf{x}_0 = 0$, then using the matrix D_k^T and (2.1.8), it follows from the first $n-1$ equations recursively that $\mathbf{x}_1 = \ldots = \mathbf{x}_{n-1} = 0$ and $V\mathbf{x}_n = 0$. By $V^+V = I$, the last equation implies that $\mathbf{x}_n = 0$, thus, $\mathbf{x} = 0$, which contradicts the assumption that $\mathbf{x}$ is an eigenvector. Let us assume that $\mathbf{x}_0 = 1 = \mathbf{P}_0$. It is readily seen that $\mathbf{x}_0,\ldots,\mathbf{x}_{n-1}$, $V\mathbf{x}_n$, and $\mathbf{P}_0,\ldots,\mathbf{P}_n$, respectively, satisfy the same n equations in the three-term relation; so are the $\mathbf{y}_0,\ldots,\mathbf{y}_n$, where $\mathbf{y}_k = \mathbf{P}_k(\Lambda) - \mathbf{x}_k$, $1 \le k \le n-1$, and $\mathbf{y}_n = V\mathbf{x}_n - \mathbf{P}_n(\Lambda)$. Since $\mathbf{y}_0 = 0$, it follows from the previous argument that $\mathbf{y}_k = 0$ for all $1 \le k \le n$, which means that $\mathbf{x}_k = \mathbf{P}_k(\Lambda)$, $1 \le k \le n-1$, and $V\mathbf{x}_n = \mathbf{P}_n(\Lambda)$. From the last equation we obtain

$$VV^+\mathbf{P}_n(\Lambda) = VV^+V\mathbf{x}_n = V\mathbf{x}_n = \mathbf{P}_n(\Lambda). \tag{4.1.16}$$

Since $U^TV = 0$, we have $U^T\mathbf{P}_n(\Lambda) = 0$. Moreover, by $V^+V = I$, we have $\mathbf{x}_n = V^+\mathbf{P}_n(\Lambda)$. The last equation in (4.1.15) then reads

$$V^+A^{*T}_{n-1,i}\mathbf{P}_{n-1}(\Lambda) + V^+B^*_{n,i}VV^+\mathbf{P}_n(\Lambda) = \lambda_i V^+\mathbf{P}_n(\Lambda).$$

Multiplying the above equation by V and using (4.1.16), the right-hand side becomes $\lambda_i\mathbf{P}_n(\Lambda)$, which we rewrite according to the three-term relation and conclude, after rearranging the terms,

$$(VV^+A^{*T}_{n-1,i} - A^T_{n-1,i})\mathbf{P}_{n-1}(\Lambda) + (VV^+B^*_{n,i} - B_{n,i})VV^+\mathbf{P}_n(\Lambda) = A_{n,i}\mathbf{P}_{n+1}(\Lambda).$$

Using (2.1.8), we obtain from the above equation that

$$\begin{aligned}\mathbf{P}_{n+1}(\Lambda) + &\sum_{i=1}^{d} D^T_{n,i}(B_{n,i} - VV^+B^*_{n,i})VV^+\mathbf{P}_n(\Lambda)\\ &+ \sum_{i=1}^{d} D^T_{n,i}(A^T_{n-1,i} - VV^+A^{*T}_{n-1,i})\mathbf{P}_n(\Lambda) = 0.\end{aligned}$$

Using (4.1.2) and (4.1.3) as well as (4.1.16), we can rewrite the above equation as

$$\begin{aligned}\mathbf{P}_{n+1}(\Lambda) + &\left(\sum_{i=1}^{d} D^T_{n,i}(I - VV^+)B^*_{n,i}VV^+ + \Gamma_1\right)\mathbf{P}_n(\Lambda)\\ &+ \left(\sum_{i=1}^{d} D^T_{n,i}(I - VV^+)A^{*T}_{n-1,i} + \Gamma_2\right)\mathbf{P}_{n-1}(\Lambda) = 0.\end{aligned}$$

Therefore, in order to conclude that $\mathbf{Q}_{n+1}(\Lambda) = 0$, we only need to show that

$$\sum_{i=1}^{d} D^T_{n,i}(I - VV^+)B^*_{n,i}VV^+ = 0, \quad \text{and} \quad \sum_{i=1}^{d} D^T_{n,i}(I - VV^+)A^{*T}_{n-1,i} = 0. \tag{4.1.17}$$

If we write $Y_i = (VV^T - I)A^T_{n-1,i}$, then by Lemma 2.3.3, (4.1.6) is equivalent to the fact that $Y = (Y_1^T, \ldots, Y_d^T)^T$ belongs to the null space of $I - A_n D_n^T$ which leads to

$$(VV^T - I)A^T_{n-1,i} = A_{n,i} \sum_{i=1}^{d} D^T_{n,i}(VV^T - I)A^T_{n-1,i} = -A_{n,i}\Gamma_2, \quad 1 \le i \le d,$$

using the formula for Γ_2 in (4.1.5). Rewriting this equation, we obtain

$$\text{(4.1.18)} \qquad VV^T A^T_{n-1,i} = A^{*T}_{n-1,i}, \quad 1 \le i \le d.$$

In particular, by $V^+V = I$, this implies that

$$(I - VV^+)A^{*T}_{n-1,i} = (I - VV^+)VV^T A^T_{n-1,i} = (V - V)V^T A^T_{n-1,i} = 0,$$

which shows that the second equation in (4.1.17) is valid. By $V^T(V^+)^T = I$, condition (4.1.7) implies

$$\text{(4.1.19)} \qquad B^*_{n,i}V = VV^T B^*_{n,i}(V^+)^T, \quad 1 \le i \le d.$$

Therefore, similar to the handling of the matrix $A^*_{n-1,i}$ above, we have

$$(I - VV^+)B^*_{n-1,i}V = (I - VV^+)VV^T B^*_{n,i}(V^+)^T = 0,$$

which verifies the first equation in (4.1.17). Thus, we have proved that the eigenvalues of $S_{n,1}, \ldots, S_{n,d}$ are zeros of $U^T\mathbf{P}_n$ and $\mathbf{Q}_{n+1}$. Next we prove that the matrices have the maximal number of real joint eigenvalues.

First we verify that each matrix $S_{n,i}$ is symmetric under the given conditions. Clearly, we only need to verify (4.1.9) and (4.1.10), but these conditions follow from (4.1.18) and (4.1.19) by using $V^+V = I$. Since all eigenvalues of symmetric matrices are real-valued, we conclude that all zeros of Q are real. Moreover, it is well-known that symmetric matrices are diagonalizable, and a family of diagonalizable matrices is simultaneously diagonalizable if, and only if, it is commuting [9, p. 52]. To show that $S_{n,1}, \ldots, S_{n,d}$ are commuting under our conditions, we only need to verify (4.1.11), (4.1.12), and (4.1.13). By (2.1.7), (4.1.11) is equivalent to (4.1.6). By the definition of $B^*_{n,i}$ in (4.1.3) and by (2.1.6), it follows that (4.1.12) is equivalent to

$$A_{n-1,i}A_{n,j}V = A_{n-1,j}A_{n,i}V$$

which, by (2.1.5), holds true. Finally, from (4.1.19) we get $U^T B^*_{n,i} V = 0$, which by (2.3.1) implies equation (4.1.14); hence (4.1.8) becomes (4.1.13) upon using (4.1.7). Therefore, our assumptions (4.1.5), (4.1.6), and (4.1.7) imply that $S_{n,1}, \ldots S_{n,d}$ are symmetric and commuting, thus, simultaneously diagonalizable. Hence, $S_{n,1}, \ldots,$ $S_{n,d}$ have $\dim \Pi^d_{n-1} + \sigma$ many real joint eigenvalues which are common zeros of $U^T \mathbf{P}_n$ and $\mathbf{Q}_{n+1}$. That these zeros are pairwise distinct follows from the fact that at each zero at least one of the partial derivatives of $U^T \mathbf{P}_n$ and $\mathbf{Q}_{n+1}$ does not vanish. This fact is a consequence of the modified Christoffel-Darboux formula, which will be proved in Section 4.2. That Q is maximal will be proved in Section 4.3 as a corollary of the uniqueness of a Lagrange interpolation problem. ∎

Remark 4.1.5. If we choose $U = I$ and $V = 0$, then Theorem 4.1.4 reduces to Theorem 1.1. On the other hand, if we choose $U = 0$ and $V = I$, then we obtain a theorem stating necessary and sufficient conditions that $\mathbf{Q}_{n+1} = \mathbf{P}_{n+1} + \Gamma_1 \mathbf{P}_n$ has $\dim \Pi^d_n$ many distinct and real zeros. Actually, this is a statement proved in [40], which leads to a minimal cubature formula of degree $2n$, see Section 8.

Remark 4.1.6. Up to now the theorem was only partially known for $d = 2$ and the cases where Möller's lower bound (1.2.10) or the bound added by one are attained; these special cases impose the restriction $\Gamma_1 = 0$ in Theorem 4.1.4, as we will show in Section 5.2. It will become evident from our development and the examples in the sections below that the presence of Γ_1 in Theorem 4.1.4 as well as the degree of freedom on the rank of V are both important.

4.2. A Christoffel-Darboux formula

In this section we establish a modified Christoffel-Darboux formula for $U^T \mathbf{P}_n$ and $\mathbf{Q}_{n+1}$. One interesting consequence of the formula will be an explicit formula for the Lagrange interpolation defined and studied in Section 4.3. For simplicity, we set

$$\mathbf{K}^{(0)}_n(\mathbf{x}, \mathbf{y}) = \mathbf{K}_n(\mathbf{x}, \mathbf{y}) + [V^+ \mathbf{P}_n(\mathbf{x})]^T V^+ \mathbf{P}_n(\mathbf{y}), \tag{4.2.1}$$

Theorem 4.2.1. *Suppose the matrices* $S_{n,1}, \ldots S_{n,d}$ *are symmetric; i.e.,* $B_{n,i}$ *and*

$A^{*T}_{n-1,i}$ *satisfy (4.1.9) and (4.1.10). Let* $\mathbf{x}, \mathbf{y} \in \mathbb{R}^d$. *If* $\mathbf{x} \neq \mathbf{y}$, *then for* $1 \leq i \leq d$,

$$(4.2.2) \quad \mathbf{K}_n^{(0)}(\mathbf{x},\mathbf{y}) = \\ \frac{(U^T\mathbf{P}_n(\mathbf{y}))^T U^+ A^T_{n-1,i}\mathbf{P}_{n-1}(\mathbf{x}) - (U^T\mathbf{P}_n(\mathbf{x}))^T U^+ A^T_{n-1,i}\mathbf{P}_{n-1}(\mathbf{y})}{x_i - y_i} \\ + \frac{(V^+\mathbf{P}_n(\mathbf{y}))^T V^+ A_{n,i}\mathbf{Q}_{n+1}(\mathbf{x}) - (V^+\mathbf{P}_n(\mathbf{x}))^T V^+ A_{n,i}\mathbf{Q}_{n+1}(\mathbf{y})}{x_i - y_i} \\ + \frac{(V^+\mathbf{P}_n(\mathbf{y}))^T V^+ B^*_{n,i}(U^+)^T U^T\mathbf{P}_n(\mathbf{x}) - (V^+\mathbf{P}_n(\mathbf{x}))^T V^+ B^*_{n,i}(U^+)^T U^T\mathbf{P}_n(\mathbf{y})}{x_i - y_i};$$

and if $\mathbf{x} = \mathbf{y}$, $1 \leq i \leq d$,

$$(4.2.3) \quad \mathbf{K}_n^{(0)}(\mathbf{x},\mathbf{x}) = (\partial_i U^T\mathbf{P}_n(\mathbf{x}))^T U^+ A^T_{n-1,i}\mathbf{P}_{n-1}(\mathbf{x}) \\ -(U^T\mathbf{P}_n(\mathbf{x}))^T U^+ A^T_{n-1,i}\partial_i\mathbf{P}_{n-1}(\mathbf{x}) + (V^+\mathbf{P}_n(\mathbf{x}))^T V^+ A_{n,i}\partial_i\mathbf{Q}_{n+1}(\mathbf{x}) \\ -(\partial_i V^+\mathbf{P}_n(\mathbf{x}))^T V^+ A_{n,i}\mathbf{Q}_{n+1}(\mathbf{x}) + (V^+\mathbf{P}_n(\mathbf{x}))^T V^+ B^*_{n,i}(U^+)^T\partial_i(U^T\mathbf{P}_n)(\mathbf{x}) \\ -(\partial_i V^+\mathbf{P}_n(\mathbf{x}))^T V^+ B^*_{n,i}(U^+)^T U^T\mathbf{P}_n(\mathbf{x}),$$

where $\partial_i = \partial/\partial x_i$ *denotes the partial derivative with respect to* x_i.

Proof. By the definition of $\mathbf{Q}_{n+1}$ in (4.3) we can rewrite the $(n+1)$-st equation in the three-term relation as

$$A^{*T}_{n-1,i}\mathbf{P}_{n-1} + B^*_{n,i}\mathbf{P}_n + A_{n,i}\mathbf{Q}_{n+1} = x_i\mathbf{P}_n,$$

which holds for every $\mathbf{x} \in \mathbb{R}^d$. Multiplying by V^+ and using (4.1.9), we obtain

$$(4.2.4) \qquad V^T A^T_{n-1,i}\mathbf{P}_{n-1} + V^+ B^*_{n,i}\mathbf{P}_n + V^+ A_{n,i}\mathbf{Q}_{n+1} = x_i V^+\mathbf{P}_n.$$

Using this equation, we have

$$(4.2.5) \quad (V^+\mathbf{P}_n(\mathbf{y}))^T V^+ A_{n,i}\mathbf{Q}_{n+1}(\mathbf{x}) - (V^+\mathbf{P}_n(\mathbf{x}))^T V^+ A_{n,i}\mathbf{Q}_{n+1}(\mathbf{y}) \\ =(x_i - y_i)(V^+\mathbf{P}_n(\mathbf{y}))^T(V^+\mathbf{P}_n)(\mathbf{x}) \\ -[(V^+\mathbf{P}_n(\mathbf{y}))^T V^+ B^*_{n,i}\mathbf{P}_n(\mathbf{x}) - (V^+\mathbf{P}_n(\mathbf{x}))^T V^+ B^*_{n,i}\mathbf{P}_n(\mathbf{y})] \\ -[(V^+\mathbf{P}_n(\mathbf{y}))^T V^T A^T_{n-1,i}\mathbf{P}_{n-1}(\mathbf{x}) - (V^+\mathbf{P}_n(\mathbf{x}))^T V^T A^T_{n-1,i}\mathbf{P}_{n-1}(\mathbf{y})].$$

Since each term in the above equation, for example $(V^+\mathbf{P}_n(\mathbf{y}))^T V^+ A_{n,i}\mathbf{Q}_{n+1}(\mathbf{x})$, is a quadratic form, its transpose is equal to itself. The first square bracket in (4.2.5), upon inserting (2.1.2) and using (4.1.10) to cancel one of the terms, can be rewritten as

$$(V^+\mathbf{P}_n(\mathbf{y}))^T V^+ B^*_{n,i}(U^+)^T U^T\mathbf{P}_n(\mathbf{x}) - (V^+\mathbf{P}_n(\mathbf{x}))^T V^+ B^*_{n,i}(U^+)^T U^T\mathbf{P}_n(\mathbf{y}).$$

Similarly, using (2.3.1) we obtain for the second square bracket in (4.2.5)

$$\begin{aligned}&[\mathbf{P}_n^T(\mathbf{y})A^T_{n-1,i}\mathbf{P}_{n-1}(\mathbf{x}) - \mathbf{P}_n^T(\mathbf{x})A^T_{n-1,i}\mathbf{P}_{n-1}(\mathbf{y})]\\ &\quad -[(U^T\mathbf{P}_n(\mathbf{y}))^T U^+ A_{n-1,i}\mathbf{P}_{n-1}(\mathbf{x}) - (U^+\mathbf{P}_n(\mathbf{x}))^T U^+ A^T_{n-1,i}\mathbf{P}_{n-1}(\mathbf{y})],\end{aligned}$$

where, using the Christoffel-Darboux formula in (2.1.11), the first term can be written as $(y_i - x_i)\mathbf{K}_n(\mathbf{x},\mathbf{y})$. Putting these equations back into (4.2.5) and dividing by $x_i - y_i$ proves (4.2.2).

To prove (4.2.3) we use the scheme such as

$$\begin{aligned}(U^T\mathbf{P}_n(\mathbf{y}))^T U^+ A_{n-1,i}\mathbf{P}_{n-1}(\mathbf{x}) &- (U^T\mathbf{P}_n(\mathbf{x}))^T U^+ A_{n-1,i}\mathbf{P}_{n-1}(\mathbf{y})\\ &=(U^T\mathbf{P}_n(\mathbf{y}) - U^T\mathbf{P}_n(\mathbf{x}))^T U^+ A_{n-1,i}\mathbf{P}_{n-1}(\mathbf{x})\\ &\quad -(U^T\mathbf{P}_n(\mathbf{x}))^T U^+ A_{n-1,i}(\mathbf{P}_{n-1}(\mathbf{y}) - \mathbf{P}_{n-1}(\mathbf{x}))\end{aligned}$$

to each term in the nominators of (4.2.2) and then let $y_i \to x_i$. ∎

If Λ is a zero of $\mathcal{Q} = \{U^T\mathbf{P}_n, \mathbf{Q}_{n+1}\}$ and at least one partial derivative of $U^T\mathbf{P}_n$ or $\mathbf{Q}_{n+1}$ at Λ is unequal to zero, then we say that Λ is a simple zero of $\mathcal{Q}$. A very useful corollary for our study is the following:

Corollary 4.2.2. *Let the assumptions be as in Theorem 4.2.1. If* $\mathbf{x}_k$ *and* $\mathbf{x}_j$ *are two zeros of* $\mathcal{Q}$*, then*

(4.2.6) $$\mathbf{K}_n^{(0)}(\mathbf{x}_k,\mathbf{x}_j) = 0, \qquad \mathbf{x}_k \neq \mathbf{x}_j,$$

and

(4.2.7) $$\begin{aligned}\mathbf{K}_n^{(0)}(\mathbf{x}_k,\mathbf{x}_k) =&(\partial_i U^T\mathbf{P}_n(\mathbf{x}))^T U^+ A^T_{n-1,i}\mathbf{P}_{n-1}(\mathbf{x})\\ &+(V^+\mathbf{P}_n(\mathbf{x}))^T V^+[A_{n,i}\partial_i\mathbf{Q}_{n+1}(\mathbf{x}) + B^*_{n,i}(U^+)^T\partial_i U^T\mathbf{P}_n(\mathbf{x})].\end{aligned}$$

Moreover, all zeros of $\mathcal{Q}$ are simple.

Proof. Clearly the two equations follow immediately from Theorem 4.2.1 by taking $\mathbf{x} = \mathbf{x}_k$ and $\mathbf{y} = \mathbf{x}_j$. By the definition of $\mathbf{K}_n^{(0)}(\cdot,\cdot)$ it is evident that $\mathbf{K}_n^{(0)}(\mathbf{x},\mathbf{x}) > 0$ for every $\mathbf{x}$, in particular for $\mathbf{x}_k$. Therefore, (4.2.7) implies that $\partial_i(U^T\mathbf{P}_n)$ and $\partial_i\mathbf{Q}_{n+1}$ cannot be both zero. Thus, $\mathbf{x}_k$ is simple. ∎

Since every zero of $\mathcal{Q} = \{U^T\mathbf{P}_n, \mathbf{Q}_{n+1}\}$ is simple, and since $\mathcal{Q}$ can have at most $\dim \Pi_{n-1}^d + \sigma$ many zeros according to Theorem 4.1.2, we conclude that if $\mathcal{Q}$ has these many zeros, then all have to be pairwise distinct. In particular, we have

Corollary 4.2.3. *Let conditions (4.1.5) to (4.1.8) in Theorem 4.1.3 be satisfied. Then $U^T\mathbf{P}_n$ and $\mathbf{Q}_{n+1}$ have $\dim \Pi_{n-1}^d + \sigma$ many real, pairwise distinct zeros.*

This closed one gap left in the proof of Theorem 4.1.3. The formulae developed above will also help us to study a special Lagrange interpolation problem in the next section.

4.3. Lagrange interpolation

We assume that a full set of zeros of $U^T\mathbf{P}_n$ and $\mathbf{Q}_{n+1}$ exists, and consider the polynomial interpolation based on these zeros. Again, we denote these zeros by $\mathbf{x}_1, \ldots, \mathbf{x}_N$, where $N = \dim \Pi_{n-1}^d + \sigma$. By Lagrange interpolation problem we mean as usual the polynomial solution of

$$P(\mathbf{x}_k) = f(\mathbf{x}_k), \qquad 1 \le k \le N, \tag{4.3.1}$$

for any given function f. However, to make this question well-posed, we need to specify a polynomial subspace from which solutions of (4.3.1) will be chosen. This subspace clearly should be a subspace of Π_n^d but it should not contain components of $U^T\mathbf{P}_n$ or any polynomial which vanishes on all zeros of $U^T\mathbf{P}_n$. The following lemma is of interests in this respect.

Lemma 4.3.1. *If $\mathcal{Q} = \{U^T \mathbf{P}_n, \mathbf{Q}_{n+1}\}$ have $\dim \Pi_{n-1}^d + \sigma$ many zeros, then any polynomial spanned by $V^+ \mathbf{P}_n$ does not vanish on all zeros of $\mathcal{Q}$.*

Proof. Suppose otherwise, then there exists an $\mathbf{a} \in \mathbb{R}^\sigma$, $\mathbf{a} \neq 0$, such that $P = \mathbf{a}^T V^+ \mathbf{P}_n$ vanishes on all zeros of $\mathcal{Q}$. Let us consider a matrix U_* defined by $U_* = ((\mathbf{a}^T V^+)^T, U)$; *i.e.*, we take $\mathbf{a}^T V^+$ as one column vector and conjunct it with U to form the new matrix. Since $U^T V = 0$, we have that

$$U_* : r_n^d \times r_n^d - \sigma + 1, \qquad \operatorname{rank} U_* = r_n^d - \sigma + 1.$$

Moreover, from our assumption, the polynomials in $\mathcal{Q}_* = \{U_*^T \mathbf{P}_n, \mathbf{Q}_{n+1}\}$ have $\dim \Pi_{n-1}^d + \sigma$ many zeros, which contradicts Theorem 4.1.2 by which these polynomials can have at most $\dim \Pi_{n-1}^d + \sigma - 1$ many zeros. The contradiction shows that $\mathbf{a}^T V^+ \mathbf{P}_n$ cannot vanish on all zeros of $\mathcal{Q}$. ∎

Because of this, we can select the subspace for which the interpolation problem is meaningfully defined by

$$\mathcal{V}_n^d = \Pi_{n-1}^d \cup \operatorname{span}\{V^+ \mathbf{P}_n\}. \tag{4.3.2}$$

We call P a Lagrange interpolation of f based on the zeros of $\mathcal{Q}$ if $P \in \mathcal{V}_n^d$ and if it satisfies (4.3.1).

From Corollary 4.2.2, we immediately have that the function $L_n(f, \cdot)$, defined by

$$L_n(f, \mathbf{x}) = \sum_{k=1}^{N} f(x_{k,n}) \frac{\mathbf{K}_n^{(0)}(\mathbf{x}, \mathbf{x}_k)}{\mathbf{K}_n^{(0)}(\mathbf{x}_k, \mathbf{x}_k)} \tag{4.3.3}$$

is a Lagrange interpolation based on the zeros of $\mathcal{Q}$. A fundamental question is whether this interpolation process is unique.

Theorem 4.3.2. *Suppose $\mathcal{Q}$ has $\dim \Pi_{n-1}^d + \sigma$ real, pairwise distinct zeros. Then $L_n(f)$ is the unique polynomial in $\mathcal{V}_n^d$ which interpolates f at the zeros of $\mathcal{Q}$.*

Proof. Clearly, an equivalent statement is that there is no polynomial in $\mathcal{V}_n^d$ which vanishes on all zeros of $\mathcal{Q}$. Suppose there exists a polynomial $P \in \mathcal{V}_n^d$ such that

$P(\mathbf{x}_k) = 0$, $1 \le k \le N$. Since $V^+\mathbf{P}_n(\mathbf{x}) = V^+G_n\mathbf{x}^n + \ldots$, we can write P as

$$P(\mathbf{x}) = V^+G_n\mathbf{x}^n + \sum_{|\alpha| \le n-1} c_\alpha \mathbf{x}^\alpha.$$

Introducing a polynomial vector $\Phi(\mathbf{x})$ whose components are the vectors $1, \mathbf{x}, \ldots \mathbf{x}^{n-1}$ and $V^+G_n\mathbf{x}^n$; *i.e.*,

$$\Phi(\mathbf{x}) = (1, \mathbf{x}^T, \ldots, (\mathbf{x}^{n-1})^T, (V^+G_n\mathbf{x}^n)^T)^T,$$

the assumption $P(\mathbf{x}_k) = 0$ means that the linear system of equations

$$(\Phi(\mathbf{x}_1)|\cdots|\Phi(\mathbf{x}_N))^T Y = 0$$

has a nontrivial solution. Hence, its coefficient matrix is singular. Suppose

$$\operatorname{rank}(\Phi(\mathbf{x}_1)|\cdots|\Phi(\mathbf{x}_N)) = m;$$

we may assume that the first $m-1$ columns of this matrix are linearly independent. That is, $\Phi(\mathbf{x}_1), \ldots, \Phi(\mathbf{x}_{m-1})$ are linearly independent and there exist scalars $a_1, \ldots a_m$, not all zero, such that

$$\sum_{k=1}^{m} a_k \Phi(\mathbf{x}_k) = 0. \tag{4.3.4}$$

In terms of components of Φ, this equation reads

$$\sum_{k=1}^{m} a_k \mathbf{x}_k^\alpha = 0, \qquad |\alpha| \le n-1, \tag{4.3.5}$$

and

$$\sum_{k=1}^{m} a_k V^+G_n\mathbf{x}_k^n = 0. \tag{4.3.6}$$

In particular, the case $\alpha = 0$ in (4.3.5) means that the sum of the a_k's is zero, which implies that at least two a_j's are nonzero. We assume $a_1 \ne 0$ and $a_m \ne 0$. By $U^T\mathbf{P}_n(\mathbf{x}_k) = 0$ we have

$$U^TG_n\mathbf{x}_k^n + \sum_{j=0}^{n-1} U^TG_{n,j}\mathbf{x}_k^j = 0,$$

which, by (4.3.5), implies that

$$\sum_{k=1}^{m} a_k U^T G_n \mathbf{x}_k^n = 0.$$

Using this equation, together with (4.3.6) and the equation

$$\mathbf{x}^n = G_n^{-1}(U^+)^T U^T G_n \mathbf{x}^n + G_n^{-1} V V^+ G_n \mathbf{x}^n,$$

which follows from (2.3.1), we conclude that

$$\sum_{k=1}^{m} a_k \mathbf{x}_k^n = 0. \tag{4.3.7}$$

Since the leading terms in $\mathbb{Q}_{n+1}$ form a basis for Π_{n+1}^d/Π_n^d, there exist $\mathbf{c}_j$ such that

$$x_j V^+ G_n \mathbf{x}^n = \mathbf{c}_j^T \mathbb{Q}_{n+1} + Q, \quad Q \in \Pi_n^d,$$

from which, by using (4.3.5), (4.3.7), and $\mathbb{Q}_{n+1}(\mathbf{x}_k) = 0$, we then have

$$\sum_{k=1}^{m} a_k x_{k,i} V^+ G_n \mathbf{x}_k^n = 0, \tag{4.3.8}$$

where we have used the notation $\mathbf{x}_k = (x_{k,1}, \ldots, x_{k,d})$. Introducing the vector Φ again, (4.3.7) and (4.3.8) imply that

$$\sum_{k=1}^{m} a_k x_{k,i} \Phi(\mathbf{x}_k) = 0.$$

Multiplying equation (4.3.4) by $x_{m,i}$ and then subtracting it from the above equation, we obtain

$$\sum_{k=1}^{m-1} a_k (x_{k,i} - x_{m,i}) \Phi(\mathbf{x}_k) = 0, \quad 1 \le i \le d.$$

Since $\Phi(\mathbf{x}_1), \ldots, \Phi(\mathbf{x}_{m-1})$ are linearly independent, we have

$$a_k (x_{k,i} - x_{m,i}) = 0, \quad 1 \le k \le m-1, \quad 1 \le i \le d.$$

Since $a_1 \neq 0$, we conclude that $x_{1,i} = x_{m,i}$ for $1 \leq i \leq d$, which is to say that $\mathbf{x}_1 = \mathbf{x}_m$, a contradiction to the assumption that the points are pairwise distinct. ∎

Clearly, the proof of this theorem has very little to do with the fact that $\mathbb{P}_n$ is an orthogonal polynomial, and the theorem is applicable as long as the interpolation points, $\mathbf{x}_1, \ldots \mathbf{x}_N$, are common zeros of linearly independent polynomials $P_1, \ldots, P_{r_n^d - \sigma}$, and there exists a $\mathbb{Q}_{n+1} = \mathbb{P}_{n+1} + \ldots$ whose components vanish on all the zeros. The idea of the proof is contained in Mysovskikh [22]. It should be mentioned that the interpolation problem considered here is rather special, but the general theorems available (cf. [16]) do not seem to apply in this case. Nevertheless, this interpolation problem seems to be quite interesting, especially the explicit formula given in (4.3.3). The connection to orthogonal polynomials should make it easier to access and possibly leads to some nice properties, such as Erdös-Turán type theorems. For the case of Gaussian cubature, we refer to [37, 40].

As a consequence of this theorem we have the following corollary, which closes the last gap in the proof of Theorem 4.1.3.

Corollary 4.3.3. *If $\mathcal{Q} = \{U^T\mathbb{P}_n, \mathbb{Q}_{n+1}\}$ has $\dim \Pi_{n-1}^d + \sigma$ many real, distinct zeros, then $\mathcal{Q}$ is maximal.*

Proof. By the definition of the maximality in **3.3**, we need to prove that if $P \in \Pi_n^d$, or $P \in \Pi_{n+1}^d$, and P vanishes on all zeros of $\mathcal{Q}$, then P belongs to the subspace spanned by $U^T\mathbb{P}_n$, or $\mathcal{Q}$, respectively. By the definition of $\mathcal{V}_n^d$ in (4.3.2) we may assume that P takes the form

$$P = \mathbf{a}^T\mathbb{Q}_{n+1} + \mathbf{b}^T U^T \mathbb{P}_n + R, \qquad R \in \mathcal{V}_n^d,$$

for some $\mathbf{a} \in \mathbb{R}^{r_{n+1}^d}$ and $\mathbf{b} \in \mathbb{R}^{r_n^d - \sigma}$. However, we then have $R \in \mathcal{V}_n^d$ and R vanishing on all $\dim \Pi_{n-1}^d + \sigma$ zeros of $\mathcal{Q}$, which implies, by Theorem 4.3.2 that $R = 0$. ∎

Our main interest in studying interpolation lies in our investigation of cubature in the following section.

4.4. Cubature formula of degree $2n-1$

Assuming that Q has $N = \dim \Pi_{n-1}^d + \sigma$ many real distinct zeros, we show that there is always a cubature formula of degree $2n-1$ with all positive weights. Moreover, we derive this formula by integrating the corresponding Lagrange interpolating polynomial, just as in the case of one variable, although there are more differences than there is similarity between the interpolation problem for one and several variables.

Theorem 4.4.1. *Let $\mathcal{L}$ be a square positive linear functional. Suppose that Q has N real distinct zeros. Then $\mathcal{L}$ admits a cubature formula of degree $2n-1$ which is based on the zeros of Q.*

Proof. Applying $\mathcal{L}$ on the explicit formula of the Lagrange interpolating polynomial $L_n(f)$ of (4.3.3) we get

$$\mathcal{L}[L_n(f)] = \sum_{k=1}^{N} \Lambda_k f(\mathbf{x}_k), \quad \Lambda_k = [\mathbf{K}_n^{(0)}(\mathbf{x}_k, \mathbf{x}_k)]^{-1}, \tag{4.4.1}$$

which is a cubature formula for $\mathcal{L}$. By the uniqueness of $L_n(f)$, it follows that this formula is exact for all polynomials in $\mathcal{V}_n^d$. However, since $U^T\mathbf{P}_n$ is orthogonal to Π_{n-1}^d and vanishes on all nodes, it follows that for any polynomial P in Π_{n-1}^d

$$\mathcal{L}(U^T\mathbf{P}_n P) = 0 = \sum_{k=1}^{N} \Lambda_k U^T\mathbf{P}_n(\mathbf{x}_k)P(\mathbf{x}_k).$$

In particular, taking $P = 1$ and using the fact that $\Pi_n^d = \mathcal{V}_n^d \cup \operatorname{span} U^T\mathbf{P}_n$, we conclude that the cubature formula (4.4.1) is exact for all polynomials in Π_n^d. Since $\mathbf{Q}_{n+1}$ vanishes on all nodes and since any polynomial $Q \in \Pi_{n+1}^d$ can be written as

$$Q = \mathbf{a}^T\mathbf{Q}_{n+1} + R, \qquad R \in \Pi_n^d,$$

for some $\mathbf{a}$, we have

$$\mathcal{L}(Q) = \mathcal{L}(R) = \sum_{k=1}^{N} \Lambda_k R(\mathbf{x}_k) = \sum_{k=1}^{N} \Lambda_k Q(\mathbf{x}_k), \quad \forall Q \in \Pi_n^d. \tag{4.4.2}$$

Thus, the cubature formula (4.4.1) is exact for all polynomials in Π^d_{n+1}. By the recursion formula (2.1.9), it readily follows that any polynomial $P \in \Pi^d_{n+2}$ can be written as

$$P = \sum_{i=1}^{d} D^T_{n+1,i} x_i \mathbf{Q}_{n+1} + R, \qquad R \in \Pi^d_{n+1}.$$

Since $P(\mathbf{x}_k) = R(\mathbf{x}_k)$ for every node and $\mathbf{Q}_{n+1}$ is orthogonal to x_i, we can follow the argument used in (4.4.2) to conclude that the cubature is exact for Π^d_{n+2}. For $k \geq 2$, by writing a polynomial in Π^d_{n+k+1} in terms of sums of $x^\alpha \mathbf{Q}_{n+1}$, $|\alpha| = k$, and a polynomial in Π^d_{n+k}, we can repeat the above process to conclude that the formula is exact for Π^d_{n+k+1} as long as $\mathbf{Q}_{n+1}$ is orthogonal to $\mathbf{x}^k$. Since $\mathbf{Q}_{n+1}$ is orthogonal to Π^d_{n-2}, we finally conclude that $k \leq n-2$ and the cubature formula is exact for Π^d_{2n-1}. ∎

As an consequence of the proof of Theorem 4.4.1, we immediately have

Corollary 4.4.2. *Let the assumptions be the same as in Theorem 4.4.1. Then the cubature formula of degree $2n-1$ based on the zeros of $\mathbf{Q}$ takes the form*

$$\mathcal{I}_n(f) = \sum_{k=1}^{N} \Lambda_{k,n} f(\mathbf{x}_k), \qquad \Lambda_{k,n} = [\mathbf{K}^{(0)}_n(\mathbf{x}_k, \mathbf{x}_k)]^{-1} > 0.$$

In particular, the compact formulae in Corollary 4.2.2 can be used to compute the weights $\Lambda_{k,n}$. From Theorem 4.1.4 and Theorem 4.4.1, we conclude:

Theorem 4.4.3. *Suppose that (4.1.6), (4.1.7), and (4.1.8) are satisfied for some V and Γ_1. Then there is a cubature formula of degree $2n-1$ which is based on the common zeros of $U^T\mathbf{P}_n$ and $\mathbf{Q}_{n+1}$.*

Remark 4.4.4. The cubature formula stated in the above theorem may not be minimal. Actually, in most cases, they will not be minimal, as we will show and discuss in detail in Section 5. However, if a minimal cubature formula is generated by the zeros of polynomials of degree n and $n+1$ but not of higher degree (recall Theorem 3.2.2), then it corresponds to a particular set of solutions of (4.1.6) – (4.1.8).

Remark 4.4.5. In a very special case for $d = 2$, Theorem 4.4.3 can be traced bac to Möller ([17–19]) where he used different notation; see also [20, 25, 28] for vario other formulations, all of which are special cases of Theorem 4.4.3; see also Remar 4.1.6.

In order to construct a cubature formula of degree $2n - 1$ based on the zer of $\mathcal{Q}$, we need to solve the equations (4.1.6), (4.1.7), and (4.1.8). These equation are, however, nonlinear, they are difficult to solve in general; even worse is the fac that they may not be solvable. The latter fact is not surprising if we recall th discussion in Section 3.2; it explains why we have to deal with the more gener cases discussed in Section 7. On the other hand, there are cases for which thes equations are indeed solvable, we shall explore them in the next two sections.

5. Möller's Lower Bound for Cubature Formula

In the previous section we provided a method of constructing cubature formulae of degree $2n-1$. For a given weight function there could be many such formulae with different numbers of nodes; among them, those with the smallest possible number we call minimal formulae. However, solving the nonlinear equations derived for the nodes of a formula is usually very difficult; the only practical way of knowing when a formula is minimal seems to be comparing its number of nodes with a known lower bound. The first lower bound is given in (1.2.9) which can be established very easily. Deeper results have been discovered, mainly by Mysovskikh and Möller. There are basically two lower bounds of general nature, both in their final form are due to Möller in his 1973 thesis. The first bound is stated in (1.2.10), the way it is proved also leads to a characterization of the cubature formulae which attain it. Although Möller's characterization in [16-18] is not cleanly formulated and a little difficult to use; there have been some improvements (cf. [20, 23, 28]), most later papers have been centered around this lower bound and the method he used.

We start this section with a discussion on the lower bound (1.2.9) and prove a new necessary condition on the existence of Gaussian cubature which seems to be of independent interest. Because of its importance, and also because of the fact that a lower bound provides a theoretic bound for rank V as well, we will study the bound given in (1.2.10) and its multidimensional extension in the second subsection. We shall reprove the lower bound by using the vector-matrix notation, which allows us to state the bound in the much more transparent form as in (1.2.10). Another reason for reproving this lower bound is that it helps us to look at it from the viewpoint of our new approach. Using the results in our previous section, we are able to give a complete characterization of the case when this bound is attained, which will be the content of Section 5.3. The second lower bound of Möller holds for centrally symmetric linear functional. Comparing it to the first one, it provide the same bound for $d = 2$, but sharper ones for $d > 2$. However, Möller's proof give little information whether or not the bound will be attained. We shall give a slightl different proof in Section 5.4 to put it into perspective to our characterization.

5.1. The first lower bound

The lower bound (1.2.9) can be easily proved as follows. Suppose the numbe of nodes N of a cubature is less than $\dim \Pi_{n-1}^d$, then there exists a polynomia $P \in \Pi_{n-1}^d$ vanishing at all nodes, which implies that $\mathcal{L}(P^2) = 0$, a contradiction t the fact that $\mathcal{L}$ is square positive.

As mentioned in the introduction, if a cubature formula attains this lowe bound, then it is called Gaussian. The theorem of Mysovskikh and Theorem 1.. show that Gaussian cubature exists if, and only if,

$$A_{n-1,i}A_{n-1,j}^T = A_{n-1,j}A_{n-1,i}^T, \quad 1 \le i,j \le d. \tag{5.1.1}$$

The question whether Gaussian cubature formula exists at all was raised by Rador in 1948 [26]; for $d = 2$ an example has been constructed numerically in [24], giving a cubature formula of degree 5. On the other hand it is known that (5.1.1) does no hold for most of the classical weight functions, including those in Examples 2.1 and 2.2 in Section 2.1 (see the next section). Examples of weight functions that admi Gaussian cubature for all n have been found only recently in [3, 29]; for $d = 2$ they are the ones given in Example 2.3 of Section 2.1. In particular, the matrices $A_{n,1}^{(-\frac{1}{2})}$ and $A_{n,2}^{(-\frac{1}{2})}$ in Example 2.3 satisfy (5.1.1) for $d = 2$, which can be easily verified.

As we will show in the following section, for quasi-centrally symmetric linear functionals (5.1.1) does not hold; hence, Gaussian cubature does not exist. By definition, for a quasi-centrally symmetric linear functional the coefficient matrices of the related three-term relation satisfy

$$B_{n,i}B_{n,j} = B_{n,j}B_{n,i}, \qquad 1 \le i,j \le d.$$

Our first theorem makes this fact more precise and provides a necessary condition for the existence of Gaussian cubature.

Theorem 5.1.1. *For a Gaussian cubature formula of degree $2n-1$ to exist it is necessary that*

$$\text{rank}(B_{n,1}|\dots|B_{n,d})\Xi_{B_n} \geq r_n^d - r_{n-2}^d = \binom{n+d-2}{n} + \binom{n+d-3}{n-1}. \tag{5.1.2}$$

Moreover, for $d = 2$, equality holds in (5.1.2), namely,

$$\text{rank}(B_{n,1}B_{n,2} - B_{n,2}B_{n,1}) = 2. \tag{5.1.3}$$

Proof. By Theorem 1.1, matrices $A_{n,i}$ satisfy the equations (5.1.1). Therefore, the commuting condition (2.1.7) is reduced to

$$B_{n-1,i}B_{n-1,j} - B_{n-1,j}B_{n-1,i} = A_{n-1,j}^T A_{n-1,i} - A_{n-1,i}^T A_{n-1,j},$$

which, adopting the notation of Ξ_C, can be written as

$$(B_{n,1}|\dots|B_{n,d})\Xi_{B_n} = (A_{n-1,1}^T|\dots|A_{n-1,d}^T)\Xi_{A_{n-1}}. \tag{5.1.4}$$

Using rank condition (2.1.3), an elementary rank inequality, and Lemma 2.3.2, we have

$$\begin{aligned}\text{rank}(B_{n,1}|\dots|B_{n,d})\Xi_{B_n} &= \text{rank}(A_{n-1,1}^T|\dots|A_{n-1,d}^T)\Xi_{A_{n-1}} \\ &\geq \text{rank}\, A_{n-1} + \text{rank}\, \Xi_{A_{n-1}} - dr_{n-1} \\ &\geq r_n^2 + dr_{n-1}^2 - r_{n-2}^2 - dr_{n-1}^2 = r_n^2 - r_{n-2}^2,\end{aligned}$$

which is (5.1.2). We notice that for $d = 2$ our vector notation reduces to

$$(B_{n,1}|B_{n,2})\Xi_{B_n} = B_{n,1}B_{n,2} - B_{n,2}B_{n,1}.$$

Thus, to prove (5.1.3) we only need to prove that the rank on the left-hand side of (5.1.3) is less than or equal to 2.

For $d = 2$, the condition (2.1.5) implies that

$$(A_{n-1,1}|A_{n-1,2})\Xi_{A_n} = 0, \qquad \Xi_{A_n} = \begin{pmatrix} A_{n,2} \\ -A_{n,1} \end{pmatrix} = \begin{pmatrix} O & -I \\ I & O \end{pmatrix} A_n.$$

By (2.1.3), it is easy to see that $\operatorname{rank} \Xi_{A_n} = r_{n+1}^2$. Since $(A_{n-1,1}|A_{n-1,2})$ is of dimension $r_{n-1}^2 \times 2r_n^2$ and $r_n^2 = n+1$ for $d = 2$, we have by $\operatorname{rank} A_{n,1} = r_n^2$ that

$$\dim \ker(A_{n-1,1}|A_{n-1,2}) = 2r_n^2 - r_{n-1}^2 = r_{n+1}^2.$$

Therefore, since Ξ_{A_n} has r_{n+1}^2 columns, we conclude that the columns of Ξ_{A_n} form a basis for the null space of $(A_{n-1,1}|A_{n-1,2})$. On the other hand, the equation (5.1.1) implies that

$$(A_{n-1,1}|A_{n-1,2})\begin{pmatrix} A_{n-1,2}^T \\ -A_{n-1,1}^T \end{pmatrix} = (A_{n-1,1}|A_{n-1,2})\Xi_{A_{n-1}^T} = 0.$$

So, the columns of $\Xi_{A_{n-1}^T}$ belongs to the null space of $(A_{n-1,1}|A_{n-1,2})$. Therefore, there exists a matrix $T_n : r_{n+1}^2 \times r_{n-1}^2$ such that $\Xi_{A_{n-1}^T} = \Xi_{A_n} T_n$, which is equivalent to

$$A_{n,1}T_n = A_{n-1,1}^T, \qquad A_{n,2}T_n = A_{n-1,2}^T. \tag{5.1.5}$$

Since by the rank condition of $A_{n-1,i}$ and by (5.1.5),

$$r_{n-1}^2 = \operatorname{rank} A_{n-1,1}^T = \operatorname{rank}(A_{n,1}T_n) \leq \operatorname{rank} T_n \leq r_{n-1}^2,$$

we have that $\operatorname{rank} T_n = r_{n-1}^2$. By (5.1.1), we can rewrite the equation (2.1.7) as

$$(A_{n-1,1}^T|A_{n-1,2}^T)\Xi_{A_{n-1}} = B_{n,2}B_{n,1} - B_{n,1}B_{n,2}.$$

We can also rewrite (5.1.1) as

$$(A_{n,1}^T|A_{n,2}^T)\Xi_{A_{n-1}^T} = 0,$$

which, together with (5.1.5) implies that

$$\begin{aligned}(B_{n+1,2}B_{n+1,1} - B_{n+1,1}B_{n+1,2})T_n =& (A_{n,1}^T|A_{n,2}^T)\Xi_{A_n} T_n \\ =& (A_{n,1}^T|A_{n,2}^T)\Xi_{A_{n-1}^T} = 0.\end{aligned}$$

Thus, we have

$$\operatorname{rank}(B_{n,2}B_{n,1} - B_{n,1}B_{n,2}) \leq \dim \ker(T_n^T) \leq r_{n+1}^2 - r_{n-1}^2 = 2.$$

This concludes the proof. ∎

As an interesting application of this theorem, we consider Example 2.3 in Section 2.1. Since it is known that Gaussian cubature exists for weight functions in this example, we have that the rank condition (5.1.3) holds; *i.e.*,

$$\operatorname{rank}(B_{n,1}^{(-\frac{1}{2})}B_{n,2}^{(-\frac{1}{2})} - B_{n,2}^{(-\frac{1}{2})}B_{n,1}^{(-\frac{1}{2})}) = 2,$$

which is not apparent from the explicit representation of the matrices given in Example 2.3.

5.2. Möller's first lower bound

Recall the discussion in Section 3.2 for minimal cubature formulae, especially equation (3.2.5). Since $\sigma_{n+k} \geq 0$, we have

$$N \geq \operatorname{rank} \Psi_n \geq \dim \Pi_{n-1}^d + \sigma_0. \tag{5.2.1}$$

If we can find an estimate for σ_0, then (5.2.1) provides a lower bound for minimal cubature.

Again, let $U^T\mathbf{P}_n$ be polynomials which vanish on all nodes of a cubature formula and let $\operatorname{rank} U = r_n^d - \sigma_0$. Möller's idea of estimating σ_0 is by determining the dimension of the subspace in Π_{n+1}^d defined by

$$\mathcal{W} = \{U^T\mathbf{P}_n, \ \ x_iU^T\mathbf{P}_n, \ \ 1 \leq i \leq d\}. \tag{5.2.2}$$

We formulate the result for all $d \geq 2$; to do so, we need to recall the notation Ξ of (2.3.3) and (2.3.4), and define $(A^T)_{n-1} = (A_{n-1,1}|\dots|A_{n-1,d})^T$ which is consistent with the notation A_n of (2.1.3).

Theorem 5.2.1. *For a minimal cubature formula, a lower bound for the number of nodes is*

$$N \geq \dim \Pi_{n-1}^d + \frac{1}{d} \operatorname{rank} \left[\Xi_{A_{n-1}^T}^T (A^T)_{n-1} \right]. \tag{5.2.3}$$

Proof. Following [18], we estimate the dimension of the subspace $\mathcal{W}$. To do so we need to consider the linear relations between the polynomials $U^T\mathbf{P}_n$ and $x_iU^T\mathbf{P}_n$ which can be found through the equation

$$\sum_{i=1}^{d} \mathbf{a}_i^T x_i U^T \mathbf{P}_n = \mathbf{b}^T \mathbf{P}_n, \tag{5.2.4}$$

where a and b are vectors to be determined. Let $\aleph$ be the number of solutions of (5.2.4). Then we have the estimates

$$(d+1)(r_n^d - \sigma_0) - \aleph \le \dim \mathcal{W} \le (r_n^d - \sigma_0) + r_{n+1}^d, \tag{5.2.5}$$

where the lower bound is clear; the upper bound follows since all $x_iU^T\mathbf{P}_n$ are polynomials of degree $n+1$ and since they span a subspace of rank at most r_{n+1}^d. From (5.2.5) it follows immediately that

$$\sigma_0 \ge r_n^d - \frac{1}{d}(r_{n+1}^d + \aleph). \tag{5.2.6}$$

We now give an upper bound of $\aleph$. Using the three-term relation in (5.2.4) and comparing the coefficients of $\mathbf{P}_{n+1}$, $\mathbf{P}_n$, and $\mathbf{P}_{n-1}$ we obtain that (5.2.4) is equivalent to

$$\sum_{i=1}^{d} \mathbf{a}_i^{*T} A_{n,i} = 0 \quad \text{and} \quad \sum_{i=1}^{d} \mathbf{a}_i^{*T} A_{n-1,i}^T = 0, \tag{5.2.7}$$

where $\mathbf{a}_i^* = U\mathbf{a}_i$ and

$$\sum_{i=1}^{d} \mathbf{a}_i^{*T} B_{n,i} = \mathbf{b}^T U^T.$$

Let $\aleph^*$ be the number of solutions of equations in (5.2.7). Then it is readily seen that $\aleph \le \aleph^*$. However, (5.2.7) is equivalent to the equation

$$\mathcal{A}\begin{pmatrix} a_1^* \\ \vdots \\ a_d^* \end{pmatrix} = 0, \quad \text{where} \quad \mathcal{A} = \begin{pmatrix} A_{n,1}^T & \cdots & A_{n,d}^T \\ A_{n-1,1} & \cdots & A_{n-1,d} \end{pmatrix}.$$

Hence, we only need to estimate the rank of $\mathcal{A}$, since

$$\aleph \le \aleph^* = dr_n^d - \operatorname{rank} \mathcal{A}. \tag{5.2.8}$$

To do so, we consider the null space of $\mathcal{A}^T$. Let $X = (X_1^T, X_2^T)^T$ be a vector belonging to this null space, where $X_1 \in \mathbb{R}^{r_{n+1}^d}$ and $X_2 \in \mathbb{R}^{r_n^d}$. Then $\mathcal{A}^T X = 0$ is equivalent to

$$A_{n,i} X_1 = A_{n-1,i}^T X_2, \qquad 1 \le i \le d.$$

By (2.1.8) we have that X_1 depends on X_2. Therefore, we only need to consider X_2; by (2.1.5) and the above equation, X_2 satisfies

$$(A_{n-1,i} A_{n-1,j}^T - A_{n-1,j} A_{n-1,i}^T) X_2 = 0, \qquad 1 \le i,j \le d,$$

which is equivalent to

$$\Xi_{A_{n-1}^T}^T (A^T)_{n-1} X_2 = 0,$$

when the coefficient matrices are put in a column according to the lexicographical order of (i,j). Therefore,

$$\dim \ker \mathcal{A}^T \le \dim \ker \left[\Xi_{A_{n-1}^T}^T (A^T)_{n-1} \right],$$

which, upon examining the sizes of the matrices involved, implies that

$$\begin{aligned} \operatorname{rank} \mathcal{A} = \operatorname{rank} \mathcal{A}^T &= r_{n+1}^d + r_{n-1}^d - \dim \ker \mathcal{A} \\ &\ge r_{n+1}^d + r_{n-1}^d - \dim \ker \left[\Xi_{A_{n-1}^T}^T (A^T)_{n-1} \right] \\ &= r_{n+1}^d + \operatorname{rank} \left[\Xi_{A_{n-1}^T}^T (A^T)_{n-1} \right]. \end{aligned}$$

Putting this estimate back into (5.2.8), and then back into (5.2.6), leads to the desired result. ∎

Remark 5.2.2. For $d = 2$, the lower bound (5.2.3) is exactly (1.2.10). Moreover, when Gaussian cubature exists, by formula (1.2.6) in Theorem 1.1, which is equivalent to $\Xi_{A_{n-1}^T}^T (A^T)_{n-1} = 0$, we have that (5.2.3) coincides with (1.2.9). Without the vector-matrix notation at the time, this lower bound appeared in [17, 18] in a different form which is not as transparent as (5.2.3).

Remark 5.2.3. By considering the subspace $\mathcal{W} = \operatorname{span}\{U^T \mathbf{P}_n,\ x_i U^T \mathbf{P}_n,\ 1 \le i \le \tau\}$, where $\tau \le d$, and following the proof of the theorem, we can also obtain

(5.2.9) $$N \ge \dim \Pi_{n-1}^d + \frac{1}{\tau}(\operatorname{rank} \mathcal{A}_\tau - r_{n+1}^d),$$

as a lower bound where

$$\mathcal{A}_\tau = \begin{pmatrix} A_{n,1}^T & \cdots & A_{n,\tau}^T \\ A_{n-1,1} & \cdots & A_{n-1,\tau} \end{pmatrix}.$$

As shown in [18], for some τ the bound in (5.2.9) may be sharper than that of (5.2.3).

The reason why we state (5.2.3) rather than (5.2.9) in the theorem is because the rank of $\mathcal{A}_\tau$ can be computed only in some special cases, the rank of $\Xi_{A_{n-1}^T}^T (A^T)_{n-1}$, however, can be computed explicitly for all centrally symmetric $\mathcal{L}$, a statement which may be of independent interests.

Theorem 5.2.4. *Let $\mathcal{L}$ be quasi-centrally symmetric. Then*

$$\text{(5.2.10)} \qquad \operatorname{rank}\left[\Xi_{A_{n-1}^T}^T (A^T)_{n-1}\right] = \begin{cases} r_n^d, & \text{if } n \text{ odd} \\ r_n^d - 1, & \text{if } n \text{ even.} \end{cases}$$

Proof. We shall use induction on n, which is based on the following consideration. Let $X_{n+1} \in \mathbf{R}^{r_{n+1}^d}$ belong to the null space of $\Xi_{A_{n+1}^T}^T (A^T)_{n+1}$; *i.e.*, let

$$\text{(5.2.11)} \qquad \Xi_{A_{n+1}^T}^T (A^T)_{n+1} X_{n+1} = 0.$$

That $\mathcal{L}$ is quasi-centrally symmetric implies, by Theorem 2.2.1, that $B_{n,i}B_{n,j} = B_{n,j}B_{n,i}$. By the commuting condition (2.1.7) we then have

$$\Xi_{A_{n+1}^T}^T (A^T)_{n+1} = \Xi_{A_n}^T A_n.$$

Therefore, $A_n X_{n+1}$ also belongs to the null space of $\Xi_{A_n}^T$. Moreover, the commuting condition (2.1.5) is equivalent to the equation

$$\Xi_{A_n}^T (A^T)_{n-1} = 0,$$

which shows that the columns of $(A^T)_{n-1} : dr_n^d \times r_{n-1}^d$ belong to the null space of $\Xi_{A_n}^T$, thus, form a basis for this space. Therefore, there exists a vector $X_{n-1} \in \mathbf{R}^{r_{n-1}^d}$ such that $A_n X_{n+1} = (A^T)_{n-1} X_{n-1}$, which implies by (2.1.8) that

$$\text{(5.2.12)} \qquad X_{n+1} = D_n^T (A^T)_{n-1} X_{n-1};$$

moreover, since the commuting condition (2.1.5) also implies that

$$\Xi^T_{A^T_{n-1}} A_n = 0,$$

we immediately obtain

$$\Xi^T_{A^T_{n-1}} (A^T)_{n-1} X_{n-1} = 0. \tag{5.2.13}$$

The equations (5.2.11), (5.2.12), and (5.2.13) imply that each vector in the null space of $\Xi^T_{A^T_{n+1}} (A^T)_{n+1}$ corresponds to one vector in the null space of $\Xi^T_{A^T_{n-1}} (A^T)_{n-1}$. Because $\Xi^T_{A_0} A_0$ is a 1×1 skew symmetric matrix, it is equal to zero, which means that there is one vector in the null space of this matrix, and we conclude that all null spaces of $\Xi^T_{A^T_n} (A^T)_n$ for n even contain only one vector. For odd n we consider the case $n = 1$. We have that $\Xi^T_{A^T_1} (A^T)_1 = \Xi^T_{A_0} A_0$, which is a matrix of size $\binom{d}{2} d \times d$. Since A_0 is a $d \times d$ matrix which is invertible by (2.1.8), we have by Lemma 2.3.1 that

$$\operatorname{rank} \left[\Xi^T_{A^T_1} A_1 \right] = \operatorname{rank} \Xi^T_{A_0} = d r^d_0 = d.$$

Therefore, for odd n the null space of $\Xi^T_{A^T_n} (A^T)_n$ is empty, which means the matrix has full rank. ∎

Remark 5.2.5. For $d > 2$, the theorem is new. For $d = 2$, (5.2.10) can be written as

$$\operatorname{rank}(A_{n-1,1} A^T_{n-1,2} - A_{n-1,2} A^T_{n-1,1}) = 2 \left[\frac{n}{2} \right], \tag{5.2.14}$$

which is due to Möller [17, 18], though he used different notation; the quasi-symmetric case appeared in [39], see also [25] where the idea of the present proof appears for $d = 2$.

For $d = 2$, the rank condition (5.2.14), thus the corresponding lower bound (5.2.3), also holds for many classical non-centrally symmetric weight functions, see [2, 44]. In particular, it is not hard to verify that the weight functions in Examples 2.1 and 2.2 in Section 2.1 all satisfy this rank condition. Indeed, for both examples, we can easily verify that $A_{n-1,1}A_{n-1,2}^T - A_{n-1,2}A_{n-1,1}^T$ takes the form

$$\begin{bmatrix} 0 & -h_0 & & & & \bigcirc \\ h_0 & 0 & -h_1 & & & \\ & \ddots & \ddots & \ddots & & \\ & & h_{n-3} & 0 & -h_{n-2} \\ \bigcirc & & & h_{n-2} & 0 \end{bmatrix};$$

hence, to verify the rank condition we only need to verify that none of the h_i's is zero. For Example 2.1 this is clear since $h_i = a_i a_{n-i-2} > 0$, where the a_i's are the coefficients in (1.1.1); for Example 2.2, this is more involved, we refer to [44].

5.3. Cubature formulae attaining the lower bound

Since the lower bound (5.2.3) is a universal lower bound for all cubature formulae of degree $2n-1$, any cubature formula of degree $2n-1$ which makes (5.2.3) an equality is a minimal cubature formula. Therefore, it is important to characterize all formulae which attain the bound.

Theorem 5.3.1. *A cubature formula of degree $2n-1$ attains the lower bound (5.2.3) if, and only if, the equations (4.1.5), (4.1.6), and (4.1.7) are solvable for a matrix V with $\Gamma_1 = -E_n$, where E_n is defined in (2.1.10), and σ is equal to the integer part of $d^{-1}\,\mathrm{rank}\,\Xi^T_{A^T_{n-1}}(A^T)_{n-1}$. Moreover, the cubature formulae are generated by the zeros of $U^T\mathbf{P}_n$.*

Proof. If (4.1.6), (4.1.7), and (4.1.8) are solved under the assumed conditions on Γ_1 and σ, then by Theorem 4.1.4 and Theorem 4.4.1 a cubature formula which attains the bound in (5.2.3) exists. On the other hand, from the proof of Theorem 5.2.1, we know that if the bound (5.2.3) is attained, then we have equalities in (5.2.5). In particular, this implies

(5.3.1) $$\dim \mathrm{span}\{x_i U^T \mathbf{P}_n,\ \ 1 \le i \le d\} = r^d_{n+1}.$$

Since $x_i\mathbf{P}_n \in \Pi^d_{n+1}$, it follows from (5.3.1) that we must have

$$\mathrm{span}\{x_i U^T \mathbf{P}_n,\ \ 1 \le i \le d\} = \mathrm{span}\,\mathbf{Q}_{n+1}$$

for some quasi-orthogonal polynomial $\mathbf{Q}_{n+1}$, which is of the form of (4.3) since the formula is of degree $2n-1$. But this implies that we can write $\mathbf{Q}_{n+1}$ in terms of linear combinations of $x_i U^T \mathbf{P}_n$, which is to say that there exist matrices $H_{n,i}$: $r_{n+1}^d \times r_n^d - \sigma$, such that

$$\mathbf{Q}_{n+1} = \sum_{i=1}^{d} H_{n,i} x_i U^T \mathbf{P}_n. \tag{5.3.2}$$

Multiplying (5.3.2) by $\mathbf{P}_{n+1}^T$ and applying the linear functional $\mathcal{L}$, we derive that

$$\sum_{i=1}^{d} H_{n,i} U^T A_{n,i} = I,$$

which means that the matrix $(H_{n,1}U^T | \cdots | H_{n,d}U^T)$ is a generalized inverse of A_n, and according to the discussion given in **2.3.1**, we can choose

$$D_{n,i}^T = H_{n,i} U^T, \qquad 1 \le i \le d. \tag{5.3.3}$$

Therefore, by (5.3.3), upon multiplying $\mathbf{P}_n$ and applying $\mathcal{L}$, we obtain that

$$\Gamma_1 = \sum_{i=1}^{d} D_{n,i}^T B_{n,i}^T = -E_n.$$

In this case, $N = \dim \Pi_{n-1}^d + \sigma$; it follows from (5.2.3) that σ is equal to the integer part of $d^{-1} \operatorname{rank} \Xi_{A_{n-1}^T}^T (A^T)_{n-1}$. That the equations (4.1.6), (4.1.7), and (4.1.8) are solvable follows again from Theorem 4.1.4. Finally, (5.3.2) implies immediately that $\mathbf{Q}_{n+1}$ vanishes whenever $U^T \mathbf{P}_n$ does. Therefore, the nodes of the cubature formula are generated by the zeros of $U^T \mathbf{P}_n$. ∎

Remark 5.3.2. This theorem gives a complete characterization of the cubature formulae which attain the lower bound (5.2.3). For $d = 2$, partial results were known in the literature (cf. [18, 20, 23, 28]).

The theorem shows that the case where the lower bound (5.2.3) is being attained is very special: the cubature formula is generated by the zeros of $U^T \mathbf{P}_n$. In other words, $\mathbf{Q}_{n+1}$ makes no real contribution; moreover, we have

$$\mathbf{Q}_{n+1} = \sum_{i=1}^{d} x_i D_{n,i}^T \mathbf{P}_n = \mathbf{P}_{n+1} - E_n \mathbf{P}_n - F_n \mathbf{P}_{n-1}, \tag{5.3.4}$$

which follows from (5.3.2) and (5.3.3). Actually, central to the above proof is equation (5.3.1), which is assumed to be true, sometimes implicitly, in almost all of the previous studies (cf. [18, 20, 28]); for example, an equivalent formulation of this condition is

$$\text{(5.3.5)} \qquad \operatorname{rank} \begin{bmatrix} U^T & & \bigcirc \\ & \ddots & \\ \bigcirc & & U^T \end{bmatrix} A_n = \operatorname{rank} \begin{bmatrix} U^T A_{n,1} \\ \vdots \\ U^T A_{n,d} \end{bmatrix} = r^d_{n+1}.$$

Apart from notational differences, for $d = 2$ (5.3.5) is exactly the rank condition which appeared in [18, Satz 2]. The following theorem indicates how strong this condition indeed is.

Theorem 5.3.3. *Suppose a cubature formula of degree* $2n-1$ *exists and its nodes are generated by zeros of* $U^T\mathbf{P}_n$. *If (5.3.1) holds, then*

$$\text{(5.3.6)} \qquad \sigma \le r^d_n - \frac{1}{d} r^d_{n+1}.$$

In particular, for $d = 2$ *and* $\mathcal{L}$ *being quasi-centrally symmetric, the lower bound of (5.2.3) is attained.*

Proof. Since (5.3.1) holds, from the proof of the previous theorem it follows that (5.3.3) holds too, which immediately implies that

$$\operatorname{rank}\,(H_{n,1}|\cdots|H_{n,d}) \begin{pmatrix} U^T & & \bigcirc \\ & \ddots & \\ \bigcirc & & U^T \end{pmatrix} = \operatorname{rank} D^T_n = r^d_{n+1}.$$

Since $\operatorname{rank} U^T = r^d_n - \sigma$, the above rank equation implies that

$$r^d_{n+1} \le d(r^d_n - \sigma),$$

which gives the desired upper bound for σ. Moreover, the number of nodes of the cubature formula satisfies the lower bound (5.2.3). If $d = 2$ and $\mathcal{L}$ is quasi-centrally symmetric, then by (5.2.14) the above upper and lower bounds give

$$\frac{n}{2} \ge \sigma \ge \left[\frac{n}{2}\right],$$

which implies $\sigma = \left[\frac{n}{2}\right]$; *i.e.*, the lower bound (5.2.3) is attained. ∎

Notice, when (5.2.3) is attained, Γ_1 is determined by U and the cubature formula is generated by $U^T\mathbf{P}_n$ alone. For the weight function $W(x,y) = w(\sqrt{x^2+y^2})$ on the unit disk, it has been proved recently that for odd n the bound (5.2.3) is not attained (see [33]). Moreover, it has been speculated and supported by numerical evidence that (5.2.3) is not sharp for many weight functions; our characterization indicates that it may well be so since it is subjected to such a strong restriction as (5.3.1). However, this suggests that the presence of Γ_1 in our theorem is very important. That is, if the lower bound (5.2.3) is not sharp, then the minimal cubature formula may be based on nodes generated by $U^T\mathbf{P}_n$ and a $\mathbf{Q}_{n+1}$ which will play an active role.

If $\mathcal{L}$ is a centrally symmetric linear functional, then by Theorem 3.1.2 the origin is a common zero of $\mathbf{P}_n$ for n odd. According to Theorem 5.3.1, if $\mathcal{L}$ admits a cubature formula such that (5.2.3) is attained, then the nodes of the formula are zeros of $U^T\mathbf{P}_n$. In particular, the following corollary is evident.

Corollary 5.3.4. *Let $\mathcal{L}$ be a centrally symmetric linear functional. If n is odd, then the origin is one of the nodes of any cubature formula that attains (5.2.3).*

In [18] Möller constructed a cubature formula of degree 9 for the product Legendre weight function of two variables which attains his lower bound. His approach used the fact that, if (5.2.3) is attained, one polynomial of degree n is known which vanishes on all nodes of the cubature formula. The general statement is as follows.

Theorem 5.3.5. *Suppose Möller's lower bound (5.2.3) is attaind by a cubature formula of degree $2n-1$ and let*

$$s_n = dr_n^d - r_{n+1}^d - \operatorname{rank}\left[\Xi^T_{A^T_{n-1}}(A^T)_{n-1}\right].$$

Then there are at least s_n many linearly independent polynomials of degree n that vanish on all nodes of the cubature formula. In fact, if $\mathbf{a}^* = (\mathbf{a}_1^*, \ldots, \mathbf{a}_d^*)$ is a nonzero solution of the linear systems given in (5.2.7), then for each i, $a_i^{*T}\mathbf{P}_n$ vanishes at all nodes of the cubature formula.

Proof. If the lower bound (5.2.3) is attained, then in the proof of Theorem 5.2.1 all the inequalities on the ranks and the dimensions of the null spaces become equality. In particular, the equation (5.2.4) has s_n many linearly independent solutions in the form $\mathbf{a} = (\mathbf{a}_1, \ldots, \mathbf{a}_d)^T$ such that $\mathbf{a}_i^* = U\mathbf{a}_i$, and $\mathbf{a}^* = (\mathbf{a}_1^*, \ldots, \mathbf{a}_d^*)$ is a solution of (5.2.7). However, since $\mathbf{a}_i^*\mathbf{P}_n = \mathbf{a}_i^{*T}U^T\mathbf{P}_n$ and $U^T\mathbf{P}_n$ vanish on all nodes of the formula, $\mathbf{a}_i^*\mathbf{P}_n$ are the polynomials which vanish at all nodes. ∎

For $d = 2$ and centrally symmetric linear functionals, we can make the result more specific.

Corollary 5.3.6. *Let $d = 2$ and $\mathcal{L}$ be a centrally symmetric linear functional. Suppose Möller's lower bound (5.2.3) is attaind by a cubature formula of degree $2n-1$. Then for odd n the two polynomials $\mathbf{a}^T A_{n,1}^T \mathbf{P}_n$ and $\mathbf{a}^T A_{n,2}^T \mathbf{P}_n$ vanish at the nodes of the cubature formula, where the vector $\mathbf{a}$ is the unique solution of the linear system $(A_{n+1,1}A_{n+1,2}^T - A_{n+1,2}A_{n+1,1}^T)\mathbf{a} = 0$.*

Proof. Since $\mathcal{L}$ is centrally symmetric and $d = 2$, it follows from (5.2.14) that

$$\operatorname{rank}\left[\Xi_{A_{n-1}^T}^T (A^T)_{n-1}\right] = \operatorname{rank}(A_{n-1,1}A_{n-1,2}^T - A_{n-1,2}A_{n-1,1}^T) = \left[\frac{n}{2}\right],$$

which implies that s_n in the previous theorem takes the value

$$s_n = 2r_n^2 - r_{n+1}^2 - 2\left[\frac{n}{2}\right] = n - 2\left[\frac{n}{2}\right].$$

In particular, if n is odd, then $s_n = 1$. From the proof of Theorem 5.3.5 there is one nonzero solution $\mathbf{a}^* = (\mathbf{a}_1^*, \mathbf{a}_2^*)^T$ of (5.2.7); *i.e*,

$$A_{n,1}^T\mathbf{a}_1^* + A_{n,2}^T\mathbf{a}_2^* = 0, \quad \text{and} \quad A_{n-1,1}\mathbf{a}_1^* + A_{n-1,2}\mathbf{a}_2^* = 0.$$

By Lemma 2.1.10 in [39], the second equation implies that there exists a nonzero vector $\mathbf{a}$ such that $\mathbf{a}_1^* = A_{n,2}\mathbf{a}$ and $\mathbf{a}_2^* = -A_{n,1}\mathbf{a}$; the first equation then implies that $\mathbf{a}$ satisfies $(A_{n,1}^T A_{n,2} - A_{n,2}^T A_{n,1})\mathbf{a} = 0$. Since $\mathcal{L}$ is centrally symmetric, thus $B_{n,i} = 0$ by Theorem 2.2.1, by (2.1.7) we conclude that $\mathbf{a}$ satisfies

$$(A_{n+1,1}A_{n+1,2}^T - A_{n+1,2}A_{n+1,1}^T)\mathbf{a} = 0.$$

The rank condition (5.2.14) implies that the last equation has exactly one solution. By arguments given in the proof of Theorem 5.3.5 we have then both $\mathbf{a}^T A_{n,1}\mathbb{P}_n$ and $\mathbf{a}^T A_{n,2}\mathbb{P}_n$ vanish at the nodes of the cubature formula. ∎

Remark 5.3.7. If (5.2.3) is attained for n odd, then we can take either $A_{n,1}\mathbf{a}$ or $A_{n,2}\mathbf{a}$, or both if they are linearly independent, as column vectors of U. As example, for the product Chebyshev weight function of the second kind, for which the bound (5.3.2) is attained as we will see in Section 6.2, the vector $\mathbf{a}$ is given as $\mathbf{a} = (0,1,0,\ldots,1,0)^T$ and the two polynomials corresponding to $\mathbf{a}^T A_{n,1}^T\mathbb{P}_n$ and $\mathbf{a}^T A_{n,2}^T\mathbb{P}_n$ are given by

$$P_0^n + P_2^n + \ldots + P_{n-1}^n, \quad \text{and} \quad P_1^n + P_3^n + \ldots + P_n^n$$

(see (6.2.3) and Example 6.2.1). Clearly, they are linearly independent. In fact, for all product weight functions, these two polynomials are linearly independent.

5.4. Möller's second lower bound

The second bound holds for centrally symmetric linear functionals. To state and prove it we need further notation. Let $\mathcal{P}_{2k}^d$ be the subspace generated by even polynomials in Π_{2k}^d, and $\mathcal{P}_{2k+1}^d$ be the subspace generated by odd polynomials in Π_{2k+1}^d. Moreover, we let $\mathcal{E}_n = \{\alpha \in \mathbb{N}_0^d : \mathbf{x}^\alpha \in \mathcal{P}_n^d\}$.

Theorem 5.4.1. *Let $\mathcal{L}$ be centrally symmetric. The number of nodes of a cubature formula of degree $2n-1$ satisfies*

$$N \geq 2\dim \mathcal{P}_{n-1}^d - \begin{cases} 1 & \text{if } n \text{ is odd} \\ 0 & \text{if } n \text{ is even.} \end{cases} \tag{5.4.1}$$

Proof. Let $\mathbf{x}_1,\ldots,\mathbf{x}_N$ be the nodes of a cubature formula. We use the notation of the matrix Ψ_n in **3.2** and define, in addition,

$$\Psi_{\mathcal{E}_n} = (\mathbf{x}_k^\alpha)_{\alpha\in\mathcal{E}_n}{}_{k=1}^{N}.$$

Then, since the rows of Ψ_n can be split into two parts which are the rows of $\Psi_{\mathcal{E}_n}$ and $\Psi_{\mathcal{E}_{n-1}}$, respectively, it is evident that

$$\operatorname{rank}\Psi_n \leq \operatorname{rank}\Psi_{\mathcal{E}_n} + \operatorname{rank}\Psi_{\mathcal{E}_{n-1}}. \tag{5.4.2}$$

We claim that $\Psi_{\mathcal{E}_{n-1}}$ has full rank; *i.e.*,

$$\text{rank}\,\Psi_{\mathcal{E}_{n-1}} = \dim \mathcal{P}^d_{n-1}. \tag{5.4.3}$$

Indeed, otherwise, there exists a vector $\mathbf{a} \neq 0$ such that $\Psi^T_{\mathcal{E}_{n-1}}\mathbf{a} = 0$, which implies that

$$P_n(\mathbf{x}) := \sum_{\alpha\in\mathcal{E}_{n-1}} a_\alpha \mathbf{x}^\alpha \in \mathcal{P}^d_{n-1}$$

vanishes at all nodes. However, since the cubature formula is of degree $2n-1$, this means that $\mathcal{L}(P_n^2) = 0$, contradicting the square positivity of $\mathcal{L}$. Next we prove that (5.4.2) is actually an equality. Indeed, suppose there is a nonzero vector in the null space of Ψ^T_n, then there is a polynomial $Q \in \Pi^d_n$, which vanishes at all nodes of the cubature formula and $Q = \mathbf{b}^T\mathbf{P}_n$ for some $\mathbf{b} \neq 0$. Since $\mathcal{L}$ is centrally symmetric, we have, as in the proof of Theorem 2.2.1, that $\mathbf{P}_n$, hence Q belongs to $\mathcal{P}^d_n$. Consequently, Q can be written as

$$Q(\mathbf{x}) = \sum_{\alpha\in\mathcal{E}_n} b_\alpha \mathbf{x}^\alpha;$$

the fact that Q vanishes at all nodes implies that the nonzero vector $(b_\alpha)_{\alpha\in\mathcal{E}_n}$ belongs to the kernel of $\Psi^T_{\mathcal{E}_n}$. Therefore, we conclude that

$$\dim\ \mathcal{P}^d_n - \text{rank}\,\Psi_{\mathcal{E}_n} = \dim\ \ker\Psi^T_{\mathcal{E}_n} \geq \dim\ \ker\Psi^T_n = \dim\Pi^d_n - \text{rank}\,\Psi_n,$$

which implies, upon using (5.4.3) and $\Pi^d_n = \mathcal{P}^d_n \cup \mathcal{P}^d_{n-1}$, that

$$\text{rank}\,\Psi_n \geq \text{rank}\,\Psi_{\mathcal{E}_n} + \dim\Pi^d_n - \dim\mathcal{P}^d_n = \text{rank}\,\Psi_{\mathcal{E}_n} + \text{rank}\,\Psi_{\mathcal{E}_{n-1}}.$$

By (5.4.2) this shows equality in (5.4.2); *i.e.*,

$$\text{rank}\,\Psi_n = \text{rank}\,\Psi_{\mathcal{E}_n} + \text{rank}\,\Psi_{\mathcal{E}_{n-1}}. \tag{5.4.4}$$

Since by the considerations of Section 3.2 we know that

$$N \geq \text{rank}\,\Psi_n, \tag{5.4.5}$$

by applying (5.4.3) and (5.4.4) the desired inequality (5.4.1) will follow once we establish the following one:

$$\operatorname{rank} \Psi_{\mathcal{E}_n} \geq \dim \mathcal{P}^d_{n-1} - \begin{cases} 1 & \text{if } n \text{ is odd,} \\ 0 & \text{if } n \text{ is even,} \end{cases} \tag{5.4.6}$$

which we will do next. Clearly, we can find a hyperplane, defined by a functional h on $\mathbf{R}^d$, such that no nonzero node lies on the hyperplane; *i.e.*,

$$h(0) = 0, \qquad h(\mathbf{x}_k) \neq 0, \quad \mathbf{x}_k \neq 0, \ 1 \leq k \leq N.$$

Let $P \in \mathcal{P}^d_{n-1}$ and n be even. Since P can not vanish at all nodes, it follows that if the origin is not a node, then $hP \in \mathcal{P}^d_n$ and $h(\mathbf{x}_k)P(\mathbf{x}_k) \neq 0$ for at least one k. Therefore, hP does not correspond to a nonzero vector in the null space of $\Psi^T_{\mathcal{E}_n}$. Since this holds for all polynomials in $\mathcal{P}^d_{n-1}$ we conclude that

$$\dim \ker \Psi^T_{\mathcal{E}_n} \leq \dim \mathcal{P}^d_n - \dim \mathcal{P}^d_{n-1}. \tag{5.4.7}$$

If the origin is a node, then for $P \in \mathcal{P}^d_{n-1}$ we have $P(0) = 0$ since P is an odd polynomial; again, (5.4.7) holds. Therefore, for n even we have proved (5.4.7), which is equivalent to (5.4.6). Finally, assume that n is odd. By Theorem 3.1.2, we have that the origin is a node. Let $\mathcal{O}_{n-1}$ denote

$$\mathcal{O}_{n-1} = \{P \in \mathcal{P}^d_{n-1} : P(0) \neq 0, \ P(\mathbf{x}_k) = 0, \ \mathbf{x}_k \neq 0\}.$$

If P_1 and P_2 both belong to $\mathcal{O}_{n-1}$, then $P_2(0)P_1 - P_1(0)P_2 \in \mathcal{P}^d_{n-1}$ vanishes at all nodes, which implies $P_2(0)P_1 = P_1(0)P_2$. Therefore, $\dim \mathcal{O}_{n-1} = 1$. For $P \in \mathcal{P}^d_{n-1}/\mathcal{O}_{n-1}$, we have that $hP(\mathbf{x}_k) \neq 0$ for at least one k. Therefore, as in the proof for n even,

$$\dim \ker \Psi^T_{\mathcal{E}_n} \leq \dim \mathcal{P}^d_n - \dim \mathcal{P}^d_{n-1} + 1,$$

which is equivalent to (5.4.6) for n odd. ∎

We want to compare the lower bound in Theorem 5.2.1 with the rank equality (5.2.10) and the one in Theorem 5.4.1, for which we need to compute $\dim \mathcal{P}^d_{n-1}$.

Theorem 5.4.2. *Let $n \in \mathbf{N}_0$. If n is even, then*

$$2\dim \mathcal{P}^d_{n-1} = \dim \Pi^d_{n-1} + \sum_{k=1}^{d-1} 2^{k-d} \dim \Pi^k_{n-1}; \tag{5.4.8}$$

if n is odd, then

$$2\dim \mathcal{P}^d_{n-1} = \dim \Pi^d_{n-1} + \sum_{k=1}^{d-1} (1 - 2^{k-d}) \dim \Pi^k_{n-2} + 1. \tag{5.4.9}$$

Proof. Since $\dim \mathcal{P}^d_n + \dim \mathcal{P}^d_{n-1} = \dim \Pi^d_n$, we only need to compute, say, $\dim \mathcal{P}^d_{n-1}$ for even n. Assume $n = 2m$. Since

$$r^d_k = \binom{k+d-1}{d-1} = \dim \Pi^{d-1}_k, \qquad k \in \mathbf{N}_0,$$

it follows that

$$\dim \mathcal{P}^d_{2m-1} = \sum_{k=1}^{m} r^d_{2k-1} = \sum_{k=1}^{m} \binom{2k+d-2}{d-1}.$$

Using the well-known combinatorial identity: $\binom{t}{i} = \binom{t-1}{i} + \binom{t-1}{i-1}$, we obtain

$$\begin{aligned} 2\dim \mathcal{P}_{2m-1} &= \sum_{k=1}^{m} \left[\binom{2k+d-2}{d-1} + \binom{2k+d-3}{d-1} \right] + \sum_{k=1}^{m} \binom{2k+d-3}{d-2} \\ &= \sum_{k=0}^{2m-1} \binom{k+d-1}{d-1} + \dim \mathcal{P}^{d-1}_{2m-1} = \dim \Pi^d_{2m-1} + \dim \mathcal{P}^{d-1}_{2m-1}, \end{aligned}$$

where the last equality follows from the identity: $\dim \Pi^d_{n-1} = r^d_0 + \ldots + r^d_{n-1}$. Therefore, we can compute $2\dim \mathcal{P}_{2m-1}$ by iteration, which gives (5.4.8). ∎

Remark 5.4.3. Using Theorem 5.4.2, it is easy to verify that the lower bound in Theorem 5.4.1 is the same as the lower bound in (5.2.3) for $d = 2$, see (5.2.14), but it yields much stronger result for $d > 2$. Nevertheless, the bound (5.2.3) holds for noncentrally symmetric $\mathcal{L}$, and it seems likely that it will be attained by some $\mathcal{L}$.

The idea of the proof of Theorem 5.4.1 is essentially due to Möller [16], the advantage of the present proof is the following. If a cubature formula attains the lower bound (5.4.1), then by the proof of Theorem 5.4.1, cf. (5.4.5), we have that

$$N = \operatorname{rank} \Psi_n.$$

By the considerations given in **3.2**, this immediately implies:

Theorem 5.4.4. *If a cubature formula attains the lower bound (5.4.1), then its nodes are generated by* $\mathcal{Q} = \{U^T\mathbb{P}_n, \mathbb{Q}_{n+1}\}$.

Looking at the formulae in Theorem 5.4.2 one may be inclined to believe that cubature formulae attaining the bound (5.4.1) are generated by certain quasi-orthogonal polynomials of higher order; the theorem shows that only $U^T\mathbb{P}_n$ and $\mathbb{Q}_{n+1}$ are needed. On the other hand, following the proof of Theorem 3.2.1, it seems likely that the lower bound (5.4.1) will not be attained in general. Theorem 5.4.4, in particular, implies that the nonlinear equations in Theorem 4.1.4 have to be solvable with σ equal to the lower bound (5.4.1). An open question is whether condition (5.3.1) is satisfied for $d > 2$; if it is, then the characterization in Theorem 5.3.1 applies. In view of the complexity of common zeros of polynomials in several variables, it seems likely that for $\mathcal{Q}$ in Theorem 5.4.4, equation (5.3.1) will be satisfied only in some peculiar cases.

6. Examples

In this section we present examples which illustrate the results derived in Sections 4 and 5. Further examples will appear in Section 8.3 and Section 9.1.

All examples in this section are given for $d = 2$ and for centrally symmetric product weight functions. To prove the existence of common zeros of $U^T\mathbf{P}_n$ and $\mathbf{Q}_{n+1}$ by the characterization given in Section 4 is equivalent to solving the nonlinear matrix equations (4.1.6), (4.1.7), and (4.1.8) for V and Γ_1. We collect results on solving these equations, which may be of general interest, in Section 6.1. In the second subsection we present examples in terms of Chebyshev weight functions of the first and the second kind; these examples illuminate our results of the previous sections, in particular, the presence of Γ_1 and the essential role played by $\mathbf{Q}_{n+1}$ in $\mathcal{Q}$. In Section 6.3 we consider general centrally symmetric product weights for small n; the results demonstrate the difficulty involved in solving these nonlinear matrix equations.

6.1. Preliminaries

All our computations are done for centrally symmetric linear functional $\mathcal{L}$ on the plane. First we specify our characterization of the common zeros of $U^T\mathbf{P}_n$ and $\mathbf{Q}_{n+1}$ in this special setting.

Theorem 6.1.1. *Let $\mathcal{L}$ be a centrally symmetric linear functional on $\mathbf{R}^2$. The set $\mathcal{Q} = \{U^T\mathbf{P}_n, \mathbf{Q}_{n+1}\}$ has $\dim \Pi^2_{n-1} + \sigma$ many pairwise distinct real zeros and $\mathcal{Q}$ is maximal if, and only if,*

$$\Gamma_2 = D^T_{n,1}(I - VV^T)A^T_{n-1,1} + D^T_{n,2}(I - VV^T)A^T_{n-1,2}, \tag{6.1.1}$$

and Γ_1 and V satisfy the following conditions:

$$A_{n-1,1}(VV^T - I)A^T_{n-1,2} = A_{n-1,2}(VV^T - I)A^T_{n-1,1} \tag{6.1.2}$$

(6.1.3) $$A_{n,i}\Gamma_1 VV^T = VV^T\Gamma_1^T A_{n,i}^T, \quad i = 1,2,$$

and

(6.1.4) $$\begin{aligned} VV^T A_{n-1,1}^T A_{n-1,2} VV^T + A_{n,1}\Gamma_1 VV^T \Gamma_1^T A_{n,2}^T \\ = VV^T A_{n-1,2}^T A_{n-1,1} VV^T + A_{n,2}\Gamma_1 VV^T\Gamma_1^T A_{n,1}^T. \end{aligned}$$

If the equations (6.1.2), (6.1.3), and (6.1.4) can be solved, then we know that $U^T\mathbf{P}_n$ and $\mathbf{Q}_{n+1}$ have $\dim \Pi_{n-1}^d + \operatorname{rank} V$ many real zeros. Clearly, $\operatorname{rank} V$ is also a parameter. Since, however, the common zeros of $U^T\mathbf{P}_n$ and $\mathbf{Q}_{n+1}$ generate a cubature formula of degree $2n-1$, the number of nodes of this formula must be greater than the number of nodes of a minimal cubature formula. Therefore, the result in Section 5 provides a lower bound for σ as well. Although this holds in general, we formulate it explicitly only for the special case.

Theorem 6.1.2. *The number of zeros of $\mathcal{Q} = \{U^T\mathbf{P}_n, \mathbf{Q}_{n+1}\}$ given in Theorem 4.1.4 satisfies the lower bound (5.2.3) and (5.4.1). In particular, for quasi-centrally symmetric $\mathcal{L}$ and $d=2$,*

(6.1.5) $$\sigma \ge \left[\frac{n}{2}\right].$$

As we pointed out in Remark 4.4.5, the presence of Γ_1 in our formula is very important. This is demonstrated in the following result.

Theorem 6.1.3. *Let $d=2$ and let $\mathcal{L}$ be centrally symmetric. If $\Gamma_1 = 0$ in Theorem 6.1.1, then*

(6.1.6) $$\sigma \le \left[\frac{n}{2}\right] + 1.$$

Proof. Let us set $A_n = A_{n-1,1}^T A_{n-1,2} - A_{n-1,2}^T A_{n-1,1}$ throughout the proof. If $\Gamma_1 = 0$, then by $V^+V = I$, equation (6.1.4) is equivalent to

$$V^T A_n V = 0.$$

In particular, it means that V belongs to the null space of $V^T A_n$. Hence,

$$\sigma \le \dim \ker V^T A_n = n + 1 - \operatorname{rank} V^T A_n.$$

It follows from (5.2.14), by the use of commuting condition (2.1.7), that

$$\operatorname{rank} A_n = A_{n,1}A_{n,2}^T - A_{n,2}A_{n,1}^T = 2\left[\frac{n+1}{2}\right],$$

which implies, by using a rank inequality (cf. [7, p.13]), that

$$\operatorname{rank} V^T A_n \geq \sigma + \operatorname{rank} A - (n+1) = \sigma + 2\left[\frac{n+1}{2}\right] - (n+1).$$

These estimates prove the theorem.

The theorem shows that if $\Gamma_1 = 0$, then the rank of V has to be either $[n/2$ or $[n/2]+1$. On the other hand, based on computer experiments H. J. Schmi has reported that for $n \geq 7$ the nonlinear equations seems to have no solution fo the weight function $W(x,y) = 1$ when $\Gamma_1 = 0$. Thus, it is to be expected that th presence of Γ_1, which allows us to choose V of a different rank, will play an essentia role in future studies. See also the remark at the end of Section 6.3.

6.2. Examples: Chebyshev weight function

We recall the product orthogonal polynomials in two variables in Section 2.1. As specific examples, we consider the ultraspherical weight function

$$(6.2.1) \qquad w_\lambda(x) = \frac{\Gamma(2\lambda+1)}{\Gamma(\lambda+1/2)\Gamma(\lambda+1/2)2^{2\lambda}}(1-x^2)^{\lambda-\frac{1}{2}}, \qquad \lambda > -\frac{1}{2},$$

whose corresponding orthogonal polynomials are the ultraspherical polynomials $P_n^{(\lambda)}(x)$ (cf. [32, Chapt. 4.7]), where for $\lambda = 0$, $P_n^{(0)}$ are the Chebyshev polynomials of the first kind ([32, (4.7.8)]). The three-term relation (1.1.1) satisfied by the orthonormal polynomials $p_n = c_n^\lambda P_n^{(\lambda)}$, where c_n^λ is a normalizing constant, has the coefficients

$$(6.2.2) \qquad a_n = \left(\frac{(n+1)(n+2\lambda)}{(2n+2\lambda+2)(2n+2\lambda)}\right)^{\frac{1}{2}}, \quad b_n = 0 \quad n \geq 0.$$

Particularly simple cases are obtained for $\lambda = 0$ and $\lambda = 1$, where the polynomials are usually termed the Chebyshev polynomials of the first and the second kind, respectively, which we give explicitly as follows.

Chebyshev weight of the first kind: $w_0(x) = \pi^{-1}(1-x^2)^{-1/2}$;

$$p_0(x) = 1, \quad p_n(x) = \sqrt{2}\cos n\theta, \quad x = \cos\theta.$$

We remark that our definition of the Chebyshev polynomials of the first kind differ from the usual one by a factor $\sqrt{2}$, since we are dealing with an orthonormal system. Moreover, the corresponding coefficients a_n in (1.1.1) are simply $a_0 = \sqrt{2}/2$ and $a_n = 1/2$ for all $n \geq 1$.

Chebyshev weight of the second kind: $w_1(x) = 2\pi^{-1}(1-x^2)^{1/2}$;

$$p_n(x) = \frac{\sin(n+1)\theta}{\sin\theta}, \qquad x = \cos\theta, \quad n \geq 0.$$

Here, the coefficients a_n in (1.1.1) are simply $a_n = 1/2$ for all $n \geq 0$.

For the Chebyshev polynomial of the first and the second kind, the coefficient matrices in the three-term relations have a very simple structure. We have $B_{n,i} = 0$ and

$$\text{(6.2.3)} \quad A_{n,1} = \frac{1}{2}\begin{bmatrix} 1 & & \bigcirc & 0 \\ & \ddots & & \vdots \\ & & 1 & 0 \\ \bigcirc & & a & 0 \end{bmatrix} \quad \text{and} \quad A_{n,2} = \frac{1}{2}\begin{bmatrix} 0 & a & & \bigcirc \\ 0 & & 1 & \\ \vdots & & & \ddots \\ 0 & \bigcirc & & 1 \end{bmatrix},$$

where $a = \sqrt{2}$ for the Chebyshev weight of the first and $a = 1$ for the second kind.

For both Chebyshev weights, Möller's lower bound (5.2.3) is attained ([20, 23, 28]); we take the case w_1 as our first example to illuminate the matrix equations, where naturally $\Gamma_1 = 0$ in Theorem 6.1.1. According to Theorem 6.1.3, if $\Gamma_1 = 0$ then $\sigma \leq [\frac{n}{2}] + 1$. Our second example considers the case when $\sigma = [\frac{n}{2}] + 1$; in this case, however, we shall see that there are polynomials in $\mathcal{Q}$ which are not spanned by $x_i U^T \mathbf{P}_n$. Our third example illustrates the situation when $\Gamma_1 \neq 0$ in Theorem 6.1.1. In this example, we will take $\sigma = n$, that is, $\sigma = r_n^2 - 1$, which is as large as it can possibly be assumed for a cubature formulae of degree $2n-1$; we shall see that almost all polynomials in $\mathcal{Q}$ are spanned by polynomials of degree $n+1$.

Example 6.2.1: Consider w_1 and $\sigma = [\frac{n}{2}]$; a cubature formula of degree $2n-1$

has $\dim \Pi^2_{n-1} + [\frac{n}{2}]$ nodes. For odd n, The matrix V defined by

$$H = I - VV^T = \frac{2}{n+1}\begin{bmatrix} 1 & 1 & \dots & 1 & \frac{1-n}{2} \\ 1 & 1 & \dots & \frac{1-n}{2} & 1 \\ \vdots & \vdots & \iddots & \vdots & \vdots \\ 1 & \frac{1-n}{2} & \dots & 1 & 1 \\ \frac{1-n}{2} & 1 & \dots & 1 & 1 \end{bmatrix}$$

solves the matrix equations (6.1.2) and (6.1.4) with $\Gamma_1 = 0$. It can be verified that $I - H$ is a nonnegative definite matrix of rank $(n-1)/2$, thus, V is well-defined. Since $V^TU = 0$ we have from $I - H = VV^T$ that $(I - H)U = 0$, from which it readily follows that the matrix $U : (n+1) \times (n+3)/2$ takes the form:

$$U^T = \begin{bmatrix} 1 & \dots & 1 & 1 & \dots & 1 \\ 1 & & \bigcirc & \bigcirc & & -1 \\ & \ddots & & & \iddots & \\ \bigcirc & & 1 & -1 & & \bigcirc \end{bmatrix}.$$

The polynomials whose zeros generate the cubature formula are given by $U^T\mathbf{P}_n$, explicitly,

$$U_k(x)U_{n-k}(y) - U_{n-k}(x)U_k(y), \quad 0 \le k \le (n-1)/2$$

and the additional polynomial

$$U_n(x)U_0(y) + U_{n-1}(x)U_0(y) + \dots + U_0(x)U_n(y).$$

The common zeros of these polynomials are given explicitly in [20]. Similarly, for even n the matrix

$$H = I - VV^T = \frac{4}{n+1}\begin{bmatrix} 1 & 0 & 1 & \dots & 1 & 0 & \frac{2-n}{4} \\ 0 & 1 & 0 & \dots & 0 & \frac{2-n}{4} & 0 \\ \vdots & \vdots & \vdots & \iddots & \vdots & \vdots & \vdots \\ 0 & \frac{2-n}{4} & 0 & \dots & 0 & 1 & 0 \\ \frac{2-n}{4} & 0 & 1 & \dots & 1 & 0 & 1 \end{bmatrix}$$

solves the matrix equations (6.1.2) and (6.1.4) with $\Gamma_1 = 0$. In this case, the matrix $U : (n+1) \times (n+2)/2$ takes the form:

$$U^T = \begin{bmatrix} 1 & 0 & \dots & 1 & 0 & 1 & 0 & 1 & \dots & 0 & 1 \\ 1 & & & & \bigcirc & 0 & \bigcirc & & & & -1 \\ & 1 & & & & \vdots & & & & -1 & \\ & & \ddots & & & \vdots & & & \iddots & & \\ & & & 1 & & 0 & & -1 & & & \\ \bigcirc & & & & 1 & 0 & -1 & & & & \bigcirc \end{bmatrix}.$$

Similar results also hold for the product Chebychev polynomials of the first kind (cf. [20, 23]).

Clearly, the matrix U is not unique; for example, for any nonsingular matrix $S:(n+1)/2\times(n+1)/2$, the common zeros of $SU^T\mathbf{P}_n$ and $U^T\mathbf{P}_n$ are the same; we can replace U^T by SU^T. In particular, we can add a multiple of one row of U^T to another to get a different U^T. In Remark 5.3.7, we showed that for odd n two vectors $A_{n,1}\mathbf{a}$ and $A_{n,2}\mathbf{a}$ can be used as two columns of U, where $\mathbf{a}$ is the solution of the linear system $(A_{n+1,1}A^T_{n+1,2}-A_{n+1,2}A^T_{n+1,1})\mathbf{a}=0$. By (6.2.3) it readily follows that $\mathbf{a}=(0,1,0,\ldots,1,0)^T$; hence, $A_{n,1}\mathbf{a}=(0,1,\ldots,0,1)$ and $A_{n,2}=(1,0,\ldots,1,0)$. In particular, $\mathbf{a}^TA^T_{n,1}+\mathbf{a}^TA^T_{n,2}=(1,1,\ldots,1,1)$ appears as the first row of U^T given above; by doing linear combination of rows, we see that both these two vectors can be used as rows of a matrix U^T.

Example 6.2.2: Consider w_0 and w_1, $\sigma=[\frac{n}{2}]+1$; a cubature formula of degree $2n-1$ has $\dim\Pi^2_{n-1}+[\frac{n}{2}]+1$ nodes. Let $\mathbf{e}_1,\ldots,\mathbf{e}_\sigma$ be the standard coordinate vectors of $\mathbb{R}^\sigma$. For $n=2m-1$, $\sigma=m$, we define the matrix $V:2m\times m$ by

$$V^T=[a\mathbf{e}_1,\mathbf{e}_2,\ldots,\mathbf{e}_m,\mathbf{e}_m,\ldots,\mathbf{e}_2,a\mathbf{e}_1],$$

while for $n=2m$, $\sigma=m+1$, we define the matrix $V:2m+1\times m+1$ by

$$V^T=[a\mathbf{e}_1,\mathbf{e}_2,\ldots,\mathbf{e}_m,2\mathbf{e}_{m+1},\mathbf{e}_m,\ldots,\mathbf{e}_2,a\mathbf{e}_1]; \tag{6.2.4}$$

here $a=\sqrt{2}$ for w_0 and $a=1$ for w_1. Using the matrices in (6.2.3), it is easy to verify that the equations (6.1.2) and (6.1.4) are satisfied by V so defined and by $\Gamma_1=0$. We now compute the polynomials in Q. By $V^TU=0$, we easily derive that

$$U^T=\begin{bmatrix}1 & & \bigcirc & \bigcirc & & -1\\ & \ddots & & & \iddots & \\ \bigcirc & & 1 & -1 & & \bigcirc\end{bmatrix},\qquad n=2m-1,$$

and

$$U^T=\begin{bmatrix}1 & & \bigcirc & 0 & \bigcirc & & -1\\ & \ddots & & \vdots & & \iddots & \\ \bigcirc & & 1 & 0 & -1 & & \bigcirc\end{bmatrix},\qquad n=2m, \tag{6.2.5}$$

which implies that the components of $U^T\mathbf{P}_n$ are given by

$$aP_0^n - aP_n^n,\ P_1^n - P_{n-1}^n,\ \ldots,\ P_{m-1}^n - P_{m+1}^n.$$

To compute $\mathbf{Q}_{n+1}$ we choose the matrix $D_{n,i}$, $i = 1, 2$, to be

$$(6.2.6)\qquad D_{n,1}^T = \begin{bmatrix} 2 & & & & \bigcirc \\ & 1 & & & \\ & & \ddots & & \\ & & & 1 & \\ \bigcirc & & & & \frac{1}{a} \\ 0 & 0 & \ldots & 0 & 0 \end{bmatrix} \quad \text{and } D_{n,2}^T = \begin{bmatrix} 0 & 0 & \ldots & 0 & 0 \\ \frac{1}{a} & & & & \bigcirc \\ & 1 & & & \\ & & \ddots & & \\ & & & 1 & \\ \bigcirc & & & & 2 \end{bmatrix},$$

and use (6.1.1) to compute Γ_2, which gives

$$(6.2.7)\qquad \Gamma_2 = \begin{bmatrix} a^2-1 & 0 & \ldots & 0 & 0 \\ \bigcirc & & & & a \\ & & & 1 & \\ & & \cdot^{\cdot^{\cdot}} & & \\ & 1 & & & \\ a & & & & \bigcirc \\ 0 & 0 & \ldots & 0 & a^2-1 \end{bmatrix}.$$

Therefore, the components of $\mathbf{Q}_{n+1}$ are given by

$$\begin{aligned} &Q_0^{n+1} = P_0^{n+1} - (a^2-1)P_0^{n-1}, \quad Q_1^{n+1} = P_1^{n+1} - aP_{n-1}^{n-1}, \\ &Q_k^{n+1} = P_k^{n+1} - P_{n-k}^{n-1}, \quad 2 \le k \le n-1, \\ &Q_n^{n+1} = P_n^{n+1} - aP_0^{n-1}, \quad Q_{n+1}^{n+1} = P_{n+1}^{n+1} - (a^2-1)P_{n-1}^{n-1}. \end{aligned}$$

Using these formulae it can be verified that

$$xU^T\mathbf{P}_n = (U^T|0)\mathbf{Q}_{n+1} \quad \text{and} \quad yU^T\mathbf{P}_n = (0|U^T)\mathbf{Q}_{n+1},$$

from which readily follows that

$$\dim \operatorname{span}\{xU^T\mathbf{P}_n, yU^T\mathbf{P}_n\} = 2m, \quad n = 2m \quad \text{or} \quad n = 2m-1.$$

We conclude that for n odd there is one polynomial of degree $n+1$ in Q which is not spanned by $\{xU^T\mathbf{P}_n, yU^T\mathbf{P}_n\}$; for n even there are two polynomials of degree $n+1$ in Q which are not spanned by $\{xU^T\mathbf{P}_n, yU^T\mathbf{P}_n\}$.

Example 6.2.3: Consider w_1, $\sigma = n$; a cubature formula of degree $2n-1$ has $\dim \Pi^2_{n-1} + n$ nodes. We take

$$V = \begin{bmatrix} 1 & & \bigcirc \\ & \ddots & \\ \bigcirc & & 1 \\ 0 & \dots & 0 \end{bmatrix} \quad \text{and} \quad \Gamma_1 = \begin{bmatrix} \bigcirc & & 1 & 0 \\ & .\cdot^{\cdot} & & \vdots \\ 1 & & \bigcirc & 0 \\ 0 & \dots & 0 & 0 \\ 0 & \dots & 0 & 0 \end{bmatrix}.$$

Since it is evident that $\Gamma_1 V V^T = \Gamma_1$ and that

$$VV^T - I = \text{diag}\{1, 0, \dots, 0, -1\}$$

it is easy to verify that the equations (6.1.2), (6.1.3), and (6.1.4) are solved by V and Γ_1 so chosen. Moreover, by $U^T V = 0$ it is clear that $U = \mathbf{e}_{n+1}$. Using (6.2.6) and (6.1.1) we deduce that

$$\Gamma_2 = \begin{bmatrix} \bigcirc & \dots & \dots & \bigcirc \\ \vdots & & & \vdots \\ \bigcirc & \dots & \dots & \bigcirc \\ 0 & \dots & 0 & 0 \\ 0 & \dots & 0 & -1 \end{bmatrix}.$$

Therefore, we have $U^T \mathbf{P}_n = P^n_n = U_n(y)$ and the components of $\mathbf{Q}_{n+1}$ are given by

$$\begin{aligned} Q^{n+1}_k(x,y) =& U_{n+1-k}(x)U_k(y) + U_{k+1}(x)U_{n-k-1}(y), \quad 0 \le k \le n-1, \\ Q^{n+1}_n(x,y) =& U_1(x)U_n(y), \quad Q^{n+1}_{n+1}(x,y) = U_{n+1}(y) - U_{n-1}(y). \end{aligned}$$

Clearly, only one component of $\mathbf{Q}_{n+1}$, namely Q^{n+1}_n, is spanned by $xU^T\mathbf{P}_n$. This example explains very well that $\Gamma_1 = 0$ is too strong a restriction for the existence of cubature formulae of degree $2n-1$.

6.3. Example: product weight function

The Examples 6.2.1 and 6.2.2 may give hope that for product weight functions the matrix equations in Theorem 6.1.1 can be solved for the case $\Gamma_1 = 0$. In particular, in Example 6.2.2 the matrix V has a particular simple form which seems to warrant an extension to the product case. Suppose that μ in Example 2.1 is symmetric on a symmetric interval so that the product measure is centrally symmetric. By (2.1.12) we have,

$$A_{n-1,1}^T A_{n-1,2} - A_{n-1,2}^T A_{n-1,1}$$
$$= \begin{bmatrix} 0 & -a_0a_{n-1} & & & \bigcirc \\ a_0a_{n-1} & 0 & -a_1a_{n-2} & & \\ & \ddots & \ddots & \ddots & \\ & & a_{n-2}a_1 & 0 & -a_{n-1}a_0 \\ \bigcirc & & & a_{n-1}a_0 & 0 \end{bmatrix}.$$

If $n = 2m-1$ is odd and if we write the matrix $V : (n+1) \times (n+1)/2$ as $V^T = [V_0, V_1, \dots, V_n]$, where V_i are the column vectors in $\mathbb{R}^m$, then the equation (6.1.4) with $\Gamma_1 = 0$ becomes

$$V^T(A_{n-1,1}^T A_{n-1,2} - A_{n-1,2}^T A_{n-1,1})V$$
$$= \sum_{k=0}^{n-1} a_k a_{n-1-k} V_{k+1} V_k^T - \sum_{k=0}^{n-1} a_k a_{n-1-k} V_k V_{k+1}^T$$
$$= \sum_{k=0}^{n-1} a_k a_{n-1-k} (V_{k+1} V_k^T - V_{n-k-1} V_{n-k}^T).$$

Therefore, if we take $V_k = V_{n-k}$, then the equation (6.1.4) is satisfied; in other word, the equation (6.1.4) is satisfied by the choice

$$V = [V_0, V_1, \dots, V_m, V_m, \dots, V_1, V_0], \qquad n = 2m-1. \tag{6.3.1}$$

Moreover, the same argument shows that for even n the choice

$$V^T = [V_0, V_1, \dots, V_{m-1}, V_m, V_{m-1}, \dots, V_1, V_0], \quad n = 2m \tag{6.3.2}$$

satisfies (6.1.4) with $\Gamma_1 = 0$. Since $\Gamma_1 = 0$, to solve the matrix equations in Theorem 6.1.1 we only need to choose the vectors V_k so that (6.1.2) is satisfied. By (2.3.2) it is readily follows that (6.1.2) is equivalent to the equation

$$L_{n-1,1} G_n^{-1} (VV^T - I)(G_n^{-1})^T L_{n-1,2}^T = L_{n-1,2} G_n^{-1} (VV^T - I)(G_n^{-1})^T L_{n-1,1}^T,$$

which implies, by the definition of $L_{n,i}$, that $G_n^{-1}(VV^T - I)(G_n^{-1})^T$ is a Hankel matrix; *i.e.*, it takes the form

$$G_n^{-1}(VV^T - I)(G_n^{-1})^T = (h_{i+j}). \tag{6.3.3}$$

We remark that this fact is true for every square positive linear functional $\mathcal{L}$. For the product weight functions, the leading coefficient matrix G_n of $\mathbb{P}_n$ takes a very simple structure (see Example 2.1), namely,

$$G_n = \begin{bmatrix} \gamma_0\gamma_n & & & & \bigcirc \\ & \gamma_1\gamma_{n-1} & & & \\ & & \ddots & & \\ & & & \gamma_{n-1}\gamma_1 & \\ \bigcirc & & & & \gamma_n\gamma_0 \end{bmatrix}.$$

Unfortunately, our choice of V in (6.3.1) or (6.3.2) turns out to be too rigid for the equation (6.3.3). Indeed, for $n = 5$ it is not hard to show that for (6.3.3) to be satisfied with V as in (6.3.1), we have to have $\gamma_1\gamma_4 = \gamma_2\gamma_3$, which holds among the product ultraspherical weight functions only for the product Chebychev weight functions of the first and the second kind. This indicates the difficulty in solving the matrix equations in Theorem 6.1.1.

In general, to solve the equations (6.1.2), (6.1.3) and (6.1.4) for centrally symmetric product weight function can be a very difficult task. The weight function $w(x) = 1$ on $[-1, 1]$, whose corresponding orthogonal polynomials are the Legendre polynomials, was initially considered by Radon [26] for $n = 3$; based on his characterization, Möller solved the case for $n = 5$ and constructed a cubature formula of degree 9 with 17 nodes, which attains his lower bound (5.2.3). In his approach Möller used Corollary 5.3.6. However, such a result does not hold true for the general situation given in Theorem 6.1.1. For $n = 5$ the cases of general centrally symmetric product weight functions have been studied in [20]. Nevertheless, we shall solve the case purely from the equations (6.1.2) and (6.1.4), and we hope that the solution, given in terms of V, may be suggestive for further studies. The cases $n < 5$ are much easier (cf. [18, 20, 28]). At the end of the section, we also comment on the possibility of solutions for $n \geq 6$.

Let $n = 5$, we choose $\Gamma_1 = 0$ and $V : 6 \times 2$ as a matrix of the form

$$V^T = \begin{bmatrix} b_1 & b_2 & b_3 & c_3 & -c_2 & c_1 \\ c_1 & c_2 & c_3 & -b_3 & b_2 & -b_1 \end{bmatrix} G_5^T,$$

where b_i and c_i are numbers to be determined and G_5 is the leading coefficient of $\mathbf{P}_5$; *i.e.*,

$$G_5 = \operatorname{diag}\{\gamma_0\gamma_5, \gamma_1\gamma_4, \gamma_2\gamma_3, \gamma_3\gamma_2, \gamma_4\gamma_1, \gamma_5\gamma_0\}.$$

Elementary calculation shows that equation (6.1.2) implies that

$$b_2^2 + c_2^2 - \gamma_1^{-2}\gamma_4^{-2} = b_1 b_3 + c_1 c_3, \tag{6.3.4}$$

$$b_2 b_3 + c_2 c_3 = b_1 c_3 - b_3 c_1, \tag{6.3.5}$$

$$b_3^2 + c_3^2 - \gamma_2^{-2}\gamma_3^{-2} = b_2 c_3 - b_3 c_2, \tag{6.3.6}$$

$$b_2 c_3 - b_3 c_2 = b_2 c_1 - b_1 c_2. \tag{6.3.7}$$

In this case the matrix in (6.1.4) becomes a 2×2 skew symmetric matrix; therefore, setting it equal to zero leads to just one condition:

$$2(b_2 c_1 - b_1 c_2)\gamma_0^2\gamma_4^2 + 2(b_3 c_2 - b_2 c_3)\gamma_1^2\gamma_3^2 + (b_3^2 + c_3^2)\gamma_2^4 = 0$$

which, upon using (6.3.6) and (6.3.7), implies

$$b_2 c_1 - b_1 c_2 = \kappa := -\frac{\gamma_2^2}{\gamma_3^2(2\gamma_2^2\gamma_4^2 - 2\gamma_1^2\gamma_3^2 + \gamma_2^4)}. \tag{6.3.8}$$

Using (6.3.8), equations (6.3.6) and (6.3.7) imply that

$$b_3^2 + c_3^2 = \gamma_2^{-2}\gamma_3^{-2} + \kappa =: A, \tag{6.3.9}$$

$$b_2 c_3 - b_3 c_2 = \kappa. \tag{6.3.10}$$

We assume that $A = \gamma_2^{-2}\gamma_3^{-2} + \kappa \geq 0$, which is equivalent to the condition

$$a_3 \leq a_1, \tag{6.3.11}$$

where the a_i's are the coefficients in (1.1.1). Moreover, this condition also implies that $\kappa < 0$. Under this assumption we can use (6.3.4) and (6.3.5) to determine b_1 and c_1, indeed,

$$b_1 = \frac{b_3(b_2^2 + c_2^2 - \gamma_1^{-2}\gamma_4^{-2}) + c_3(b_2 c_3 + b_2 b_3)}{\gamma_1^{-2}\gamma_4^{-2} + \kappa},$$

$$c_1 = \frac{c_3(b_2^2 + c_2^2 - \gamma_1^{-2}\gamma_4^{-2}) - b_3(b_2c_3 + b_2b_3)}{\gamma_1^{-2}\gamma_4^{-2} + \kappa}.$$

Using these two formulae to simplify (6.3.8), we get

$$\kappa(\gamma_1^{-2}\gamma_4^{-2} + \kappa) = \kappa(b_2^2 + c_2^2 - \gamma_2^{-2}\gamma_3^{-2}) - (b_2b_3 + c_2c_3)^2.$$

We combine this equation and equation (6.3.10) and derive, after some simplifications,

$$b_2^2 + c_2^2 = -\kappa(\gamma_2^2\gamma_3^2\gamma_1^{-2}\gamma_4^{-2} + 1) := B. \tag{6.3.12}$$

Thus, we only need to find conditions on b_2, b_3 and c_2, c_3 such that (6.3.9), (6.3.10), and (6.3.12) are satisfied. These equations are solvable if $\kappa(AB)^{-1/2} \geq -1$ since this condition allows us to choose

$$b_2 = \sqrt{A}\cos\theta, \quad c_2 = \sqrt{A}\sin\theta, \quad b_3 = \sqrt{b}\cos\phi, \quad c_3 = \sqrt{B}\sin\phi,$$

where θ and ϕ are chosen such that $\sin(\theta - \phi) = \kappa(AB)^{-1/2}$. The condition $\kappa(AB)^{-1/2} \geq -1$ is equivalent to

$$\frac{2}{a_2^2}\left(\frac{1}{a_3^2} + \frac{1}{a_1^2}\right)(1 - a_1^2a_3^2) \leq 1. \tag{6.3.13}$$

To sum up we conclude that

Theorem 6.3.1. *For the product weight function, there exists a cubature formula of degree 9 with 17 nodes if (6.3.11) and (6.3.13) are satisfied.*

As an example we consider the ultraspherical polynomials. By (6.2.2) and the fact that $\lambda > -1/2$, it is easy to derive the following rough estimates: $a_1^{-2} \geq 2/3$, $a_2^{-2} \geq 5/3$, and $a_3^{-2} \geq 7/4$, from which immediately follows that (6.3.13) holds for all w_λ, $\lambda > -1/2$. However, it is easy to verify that (6.3.11) holds only if $\lambda \leq 1$.

The difficulties in solving the matrix equations for such a special case indicates that the problem is extreme difficult in general; however, this is what one expects when dealing with common zeros of a family of polynomials in several variables.

Finally, we want to comment on the possibility of solving the matrix equations for $n \geq 7$. As we mentioned at the end of Section 6.1, numerical experiments indicate that the equations (6.1.2) and (6.1.4) have no solution when $\Gamma_1 = 0$. However, when we take Γ_1 into consideration, a solution is very likely to exist. To be sure, let us compare the equations with Γ_1 and those without. The equation (6.1.3) is equivalent to (6.3.6), therefore, equation (6.1.3) leaves $2n + 1$ degrees of freedom at our disposal. On the other hand, the number of equations in (6.1.2) and (6.1.4) do not change with or without the Γ_1 term. Therefore, we see that we gain $2n + 1$ degrees of freedom by taking Γ_1 into consideration and solving (6.1.2), (6.1.3), and (6.1.4) jointly, which should increase the chance of finding a solution greatly. Since, however, these equations are nonlinear, there is no guarantee that a solution will exist even with the additional freedom, solving these equations are generally very difficult.

7. Common Zeros of Polynomials in Several Variables: General Case

In this section we will characterize the common zeros of a set of quasi-orthogonal polynomials in several variables in general; *i.e.*, we will study the set of polynomials $\mathcal{Q} = \{U_0^T \mathbf{Q}_n, \ldots, U_{r+1}^T \mathbf{Q}_{n+r+1}\}$, where we follow the definition given in Section 3.3. The characterization of the common zeros of $\mathcal{Q}$ is again given in terms of a system of nonlinear matrix equations, which will contain further equations and further unknowns which makes it even more formidable to be solved. However, all cubature formulae are based on zeros of a certain set of quasi-orthogonal polynomials; a characterization in terms of matrix equations is certainly an important step towards the construction of formulae. Moreover, our characterization is complete; *i.e.*, the existence of a cubature formula is completely converted to the solvability of matrix equations, which is an algebraic or a numerical problem; the complete characterization also makes it possible to deal with the problem by use of other methods in the future. For these reasons, we present the results in this section. Naturally, we follow the development of Section 4, and even some of the proofs are done through induction where the results in Section 4 serve as the initial step in the induction. However, to avoid redundant arguments and repetition of proofs, we shall be brief whenever possible; for some of the theorems we only provide an indication of the proof. For the central theorem in Section 7.1, however, which gives the most general characterization, a complete proof will be given; one reason for doing so is that from this general statement, further results as the ones in Section 8 will follow. One major problem which occurs in the general setting lies in the explicit formulae for various matrices which appear in the recurrence relations satisfied by $\mathbf{Q}_{n+k}$; even for a moderate size of r these formulae may become too complicated to be tackled. This difficulty will be avoid by working with a family of modified matrices; as a trade off, we will not be able to give explicit formulae for $\mathbf{Q}_{n+k}$ in general.

We will not give examples for the results obtained in the section, but an example will be given to construct a cubature formula of even degree, which from the point of view of solving nonlinear matrix equations corresponds to a special case of those given in this section (cf. (7.1.6) - (7.1.9)).

Throughout the section we follow the definitions in Section 3.3 for quasi-orthogonal polynomials $\mathbf{Q}_{n+k}$; in particular, for the definitions of the matrices U_k, V_k and Q.

7.1. Characterization of zeros

We begin with an analogy of Lemma 4.1.1, which connects the zeros of Q to joint eigenvalues of matrices.

Lemma 7.1.1. *If $\Lambda = (\lambda_1, \ldots, \lambda_d)$ is a zero of Q, then Λ is a joint eigenvalue of a family of matrices $S^{(r)}_{n,1}, \ldots, S^{(r)}_{n,d}$ which has a joint eigenvector, where $S^{(r)}_{n,i}$ is defined as*

$$S^{(r)}_{n,i} = \begin{bmatrix} B_{0,i} & A_{0,i} & & & & & \bigcirc \\ A^T_{0,i} & B_{1,i} & & & & & \\ \ddots & \ddots & \ddots & & & & \\ & & B_{n-2,i} & A_{n-1,i}V_0 & & & \\ & & V_0^+ A^{*T}_{n-1,i} & V_0^+ B^*_{n,i} V_0 & V_0^+ A_{n,i} V_1 & & \\ & & \ddots & \ddots & \ddots & & \\ & & \cdots & \ddots & V^+_{r-1} B^*_{n+r-1,i} V_{r-1} & V^+_{r-1} A_{n+r-1,i} V_r \\ \bigcirc & \cdots & \cdots & \cdots & V_r^+ A^{*T}_{n+r-1,i} V_{r-1} & V_r^+ B^*_{n+r,i} V_r \end{bmatrix};$$

we shall not need the exact formulae for the matrices in the lower left corner and for $A^*_{n+k,i}$; for $0 \le k \le r$

$$B^*_{n+k,i} = B_{n+k,i} + \Gamma_{1,k} A_{n+k-1,i} - A_{n+k,i}\Gamma_{1,k+1}. \tag{7.1.1}$$

Moreover, the joint eigenvector is given by

$$[\mathbf{P}_0^T(\Lambda), \ldots, \mathbf{P}_{n-1}^T(\Lambda), (V_0^+\mathbf{Q}_n(\Lambda))^T, \ldots, (V_r^+\mathbf{Q}_{n+r}(\Lambda))^T]^T. \tag{7.1.2}$$

Proof. Suppose Λ is a common zero of $\mathcal{Q}$. The fact that Λ is an eigenvalue of $S_{n,1}, \ldots, S_{n,d}$ is equivalent to $S_{n,i}\mathbf{x} = \lambda_i \mathbf{x}$, where $\mathbf{x}$ is given as in (7.1.2), which is equivalent to $n+r$ matrix equations. That Λ satisfies the first n equations is shown in Lemma 4.1.1. The rest of the proof amounts to rewriting the three-term relation in terms of $\mathbb{Q}_{n+k}$ and then applying the condition $U_k^T \mathbb{Q}_{n+k}(\Lambda) = 0$. Indeed, since $\mathbb{Q}_{n+k} = \mathbb{P}_{n+k} + \ldots$, it is clearly possible to use the three-term relation to write $x_i \mathbb{Q}_{n+k}$ in terms of $\mathbb{Q}_{n+k+1}, \mathbb{Q}_{n+k}, \ldots$ etc. More precisely, we have

$$\text{(7.1.3)} \quad \begin{aligned} x_i \mathbb{Q}_{n+k} =& A_{n+k,i}\mathbb{Q}_{n+k+1} + B^*_{n+k,i}\mathbb{Q}_{n+k} + A^{*T}_{n+k-1,i}\mathbb{Q}_{n+k-1} \\ &+ A^{(k)}_{n+k-2}\mathbb{Q}_{n+k-2} + \ldots + A^{(k)}_{n,i}\mathbb{P}_n + \ldots + A^{(k)}_{n-k-1,i}\mathbb{P}_{n-k-1}. \end{aligned}$$

where $A^{(k)}_{n+j}$ are proper matrices. Multiplying this equation by $\mathbb{P}_{n+r}$ and applying $\mathcal{L}$, we have that $B^*_{n+k,i}$ is of the form (7.1.1). Similarly, we can multiply the equation by $\mathbb{P}_{n+r-1}$ and derive an explicit formula for A^*_{n+k-1}, and so on. The condition $U_k^T \mathbb{Q}_{n+k} = 0$ leads, by use of (2.3.1), to

$$\mathbb{Q}_{n+k}(\Lambda) = ((U_k^+)^T U_k^T + V_k V_k^+)\mathbb{Q}_{n+k}(\Lambda) = V_k V_k^+ \mathbb{Q}_{n+k}(\Lambda).$$

Taking (7.1.3) at the point Λ and multiply the equation by V_k^+, the above relation then implies that Λ satisfies the remaining equations that are equivalent to $S_{n,i}\mathbf{x} = \lambda_i \mathbf{x}$. ∎

Theorem 7.1.2. *The set $\mathcal{Q}$ has at most $N := \dim \Pi_{n-1} + \sigma_0 + \ldots + \sigma_r$ distinct zeros.*

The proof of this theorem is similar to that of Theorem 4.1.2. Before stating the main result, we first need to state a lemma, the proof of which follows from considerations given in the proof of Lemma 4.1.3 and from the definition of maximality given in Section 3.3.

Lemma 7.1.3. *If $\mathcal{Q}$ is maximal, then for $0 \le k \le r$,*

$$\text{(7.1.4)} \qquad U_{k-1}^T A_{n+k-1,i} V_k = 0, \quad U_k^T B^*_{n+k,i} V_k = 0, \quad 1 \le i \le d.$$

where we take $U_{-1} = 0$.

The main theorem in this section, Theorem 7.1.4 below, is similar to that of Theorem 4.1.4. However, because of the complicated structure of the matrices $S_n^{(r)}$, we shall state and prove the theorem somewhat differently; instead of working with $S_n^{(r)}$, we shall work with matrices $T_n^{(r)}$, which are of block tri-diagonal form, defined as

$$T_{n,i}^{(r)} = \begin{bmatrix} B_{0,i} & A_{0,i} & & & & & \bigcirc \\ A_{0,i}^T & B_{1,i} & & & & & \\ & \ddots & \ddots & \ddots & & & \\ & & B_{n-2,i} & A_{n-1,i}V_0 & & & \\ & & V_0^T A_{n-1,i}^T & V_0^+ B_{n,i}^* V_0 & V_0^+ A_{n,i} V_1 & & \\ & & \ddots & \ddots & \ddots & & \\ & & & & V_{r-1}^+ B_{n+r-1,i}^* V_{r-1} & V_{r-1}^+ A_{n+r-1,i} V_r \\ \bigcirc & & & & V_r^T A_{n+r-1,i}^T (V_{r-1}^+)^T & V_r^+ B_{n+r,i}^* V_r \end{bmatrix}$$

for $1 \le i \le k$. The point is that if $S_{n,i}^{(r)}$ is symmetric, then it will look like $T_{n,i}^{(r)}$.

Theorem 7.1.4. *Let U_k, $0 \le k \le r$, be defined as in Section 3.3 and let $U_{r+1} = I$. Then there exist quasi-orthogonal polynomials $\mathbf{Q}_{n+k}$ of order $2k$ in the form of*

$$\mathbf{Q}_{n+k} = \mathbf{P}_{n+k} + \Gamma_{1,k}\mathbf{P}_{n+k-1} + \ldots, \quad 0 \le k \le r, \tag{7.1.5}$$

such that $Q = \{U_0^T \mathbf{Q}_n, \ldots, U_{r+1}^T \mathbf{Q}_{n+r+1}\}$, where $\mathbf{Q}_n = \mathbf{P}_n$, has $\dim \Pi_{n-1}^d + \sigma_0 + \cdots + \sigma_r$ many pairwise distinct real zeros and Q is maximal if, and only if, Γ_k and V_k satisfy the following conditions:

$$A_{n-1,i}(V_0V_0^T - I)A_{n-1,j}^T = A_{n-1,j}(V_0V_0^T - I)A_{n-1,i}^T \quad 1 \le i,j \le d, \tag{7.1.6}$$

$$B_{n+k,i}^* V_k V_k^T = V_k V_k^T B_{n+k,i}^{*T}, \quad 0 \le k \le r, \quad 1 \le i \le d, \tag{7.1.7}$$

$$U_k^T A_{n+k,i} V_{k+1} = 0 \quad 1 \le i \le d, \quad 0 \le k \le r-1, \tag{7.1.8}$$

and for $0 \le k \le r-1$, $1 \le i,j \le d$,

$$(7.1.9)\quad \begin{aligned} &V_k V_k^T A_{n+k-1,i}^T (V_{k-1}^+)^T V_{k-1}^+ A_{n+k-1,j} V_k V_k^T + B_{n+k,i}^* V_k V_k^T B_{n+k,j}^{*T} \\ &+A_{n+k,i} V_{k+1} V_{k+1}^T A_{n+k,j}^T = A_{n+k,j} V_{k+1} V_{k+1}^T A_{n+k,i}^T \\ &+V_k V_k^T A_{n+k-1,j}^T (V_{k-1}^+)^T V_{k-1}^+ A_{n+k-1,i} V_k V_k^T + B_{n+k,j}^* V_k V_k^T B_{n+k,i}^{*T}, \end{aligned}$$

where we take $V_{r+1} = 0$ and $V_{-1} = I$.

Proof. First, suppose that Q has $\dim \Pi_{n-1}^d + \sigma_0 + \cdots + \sigma_r$ many real distinct zeros. Then, by Lemma 7.1.1, the corresponding matrices $S_{n,1}^{(r)}, \ldots, S_{n,d}^{(r)}$ have these many distinct joint eigenvalues. Therefore, $S_{n,1}^{(r)}, \ldots, S_{n,d}^{(r)}$ have to be symmetric and simultaneously diagonalizable. The symmetry of $S_{n,i}$ implies that $A_{k,i}^{(j)} = 0$ and that for $0 \le k \le r$

$$(7.1.10)\qquad V_k^+ A_{n+k-1,i}^{*T} V_{k-1} = V_k^T A_{n+k-1,i}^T (V_{k-1}^+)^T, \qquad 1 \le i \le d,$$

as well as

$$(7.1.11)\qquad V_k^+ B_{n+k,i}^* V_k = V_k^T B_{n+k,i}^{*T} (V_k^+)^T, \qquad 1 \le i \le d.$$

Since Q is maximal, we obtain from (7.1.4) and (2.3.1) that (7.1.11) implies

$$(7.1.12)\qquad B_{n+k,i}^* V_k = V_k V_k^T B_{n+k,i}^{*T} (V_k^+)^T,$$

which, upon multiplying by V_k^T and using (7.1.4) and (2.3.1) again, is equivalent to (7.1.7). The equation (7.1.8) is the same as the first equation in (7.1.4). Taking (7.1.10) and (7.1.11) into consideration and using the commuting conditions (2.1.5), (2.1.6) and (2.1.7), we obtain that $S_{n,i}^{(r)}$, $1 \le i \le d$, are mutually commuting which is equivalent to

$$(7.1.13)\quad \begin{aligned} &A_{n-2,i}^T A_{n-2,j} + B_{n-1,i} B_{n-1,j} + A_{n-1,i} V V^T A_{n-1,j}^T \\ &= A_{n-2,j}^T A_{n-2,i} + B_{n-1,j} B_{n-1,i} + A_{n-1,j} V V^T A_{n-1,i}^T, \end{aligned}$$

$$(7.1.14)\quad \begin{aligned} &V_k^+ B_{n+k,i}^* V_k V_k^+ A_{n+k,j} V_{k+1} + V_k^+ A_{n+k,i} V_{k+1} V_{k+1}^+ B_{n+k+1,j}^* V_{k+1} \\ &= V_k^+ B_{n+k,j}^* V_k V_k^+ A_{n+k,i} V_{k+1} + V_k^+ A_{n+k,j} V_{k+1} V_{k+1}^+ B_{n+k+1,i}^* V_{k+1} \end{aligned}$$

for $0 \le k \le r-1$, and

$$(7.1.15)\quad V_k^T A_{n+k-1,i}^T (V_{k-1}^+)^T V_{k-1}^+ A_{n+k-1,j} V_k + V_k^+ B_{n+k,i}^* V_k V_k^T B_{n+k,j}^{*T} (V_k^+)^T$$
$$+ V_k^+ A_{n+k,i} V_{k+1} V_{k+1}^T A_{n+k,j}^T (V_k^+)^T = V_k^+ A_{n+k,j} V_{k+1} V_{k+1}^T A_{n+k,i}^T (V_k^+)^T$$
$$+ V_k^T A_{n+k-1,j}^T (V_{k-1}^+)^T V_{k-1}^+ A_{n+k-1,i} V_k + V_k^+ B_{n+k,j}^* V_k V_k^T B_{n+k,i}^{*T} (V_k^+)^T$$

for $0 \le k \le r$. By (2.1.7) it follows that (7.1.13) is equivalent to (7.1.6), while by (2.3.1) we have from (7.1.4) that

$$(7.1.16)\qquad V_{k-1} V_{k-1}^+ A_{n+k-1,i} V_k = A_{n+k-1,i} V_k, \qquad 1 \le k \le r,$$

and

$$(7.1.17)\qquad V_k V_k^+ B_{n+k,i}^* V_k = B_{n+k,i}^* V_k, \qquad 1 \le i \le d.$$

Using these two equations, it is readily seen that (7.1.15) transforms into (7.1.9). Thus, the necessity of all conditions have been verified.

On the other hand, assume that conditions (7.1.6) – (7.1.9) are satisfied. First we prove that every eigenvalue of $T_{n,1}^{(r)}, \ldots, T_{n,d}^{(r)}$ which has a joint eigenvector $\mathbf{x}$ must be a common zero of all $U_k^T \mathbf{Q}_n$, $0 \le k \le r+1$. Suppose that $\Lambda = (\lambda_1, \ldots, \lambda_d)$ is such an eigenvalue and its eigenvector is written as

$$\mathbf{x} = (\mathbf{x}_0^T, \ldots, \mathbf{x}_{n+r-1}^T)^T, \quad \mathbf{x}_k \in \mathbb{R}^{r_k^d}, \quad 1 \le k \le n-1; \quad \mathbf{x}_{n+k} \in \mathbb{R}^{\sigma_k}, \quad 0 \le k \le r-1.$$

Then, as in the proof of Theorem 4.1.4, it follows from $T_{n,i}^{(r)}\mathbf{x} = \lambda_i \mathbf{x}$ that $\{\mathbf{x}_j\}$ satisfies a three-term relation whose first n equations are exactly the equations (4.1.15) and whose other equations are

$$(7.1.18)\qquad V_k^T A_{n+k-1,i}^T (V_{k-1}^+)^T \mathbf{x}_{n+k-1} + V_k^+ B_{n+k,i}^* V_k \mathbf{x}_{n+k}$$
$$+ V_k^+ A_{n+k,i} V_{k+1} \mathbf{x}_{n+k+1} = \lambda_i \mathbf{x}_{n+k}, \quad 0 \le k \le r.$$

From the condition (7.1.8) we obtain by (2.3.1) that (7.1.16) holds. Using (7.1.16) and (2.1.8), together with (2.3.1), we derive from (7.1.18) that

$$(7.1.19)\ V_{k+1}\mathbf{x}_{n+k+1} = \sum_{i=1}^{d} D_{n+k,i}^T \lambda_i V_k \mathbf{x}_{n+k} - \sum_{i=1}^{d} D_{n+k,i}^T V_k V_k^+ B_{n+k,i}^* V_k \mathbf{x}_{n+k}$$
$$- \sum_{i=1}^{d} D_{n+k,i}^T V_k V_k^T A_{n+k-1,i}^T (V_{k-1}^+)^T \mathbf{x}_{n+k-1}.$$

Again we show that $\mathbf{x}_0$, thus $\mathbf{x}$, is not zero. Using the first n equations of the recurrence relation we have as in the proof of Theorem 4.1.4 that if $\mathbf{x}_0 = 0$ then $\mathbf{x}_1 = \ldots = \mathbf{x}_{n-1} = 0$ and $V_0\mathbf{x}_n = 0$. By $V_0^+V_0 = I$, the last equation implies that $\mathbf{x}_n = 0$. Moreover, by (7.1.19), we can use induction to conclude that $V_k\mathbf{x}_{n+k} = 0$, which implies that $\mathbf{x}_{n+k} = 0$, $0 \le k \le r-1$. But then, we get $\mathbf{x} = 0$, which contradicts the assumption that $\mathbf{x}$ is an eigenvector. Therefore, we may assume that $\mathbf{x}_0 = 1 = \mathbf{P}_0$. Again as in the proof of Theorem 4.1.4, we can conclude from (4.1.15) that $\mathbf{x}_k = \mathbf{P}_k(\Lambda)$, $1 \le k \le n-1$, and $V_0\mathbf{x}_n = \mathbf{P}_n(\Lambda)$; moreover, it follows that $U_0^T\mathbf{P}_n(\Lambda) = 0$ and $\mathbf{x}_n = V_0^+\mathbf{P}_n(\Lambda)$. To conclude that there exist $\mathbf{Q}_{n+k}$ in the form of (7.1.5) such that $U_k^T\mathbf{Q}_{n+k}(\Lambda) = 0$, we use induction based on (7.1.19). Assume that we have proved that there exist $\mathbf{Q}_{n+j}$ which are quasi-orthogonal polynomials of order $2j$ such that

$$\mathbf{x}_{n+j} = V_j^+\mathbf{Q}_{n+j}(\Lambda), \quad \text{and} \quad U_j^T\mathbf{Q}_{n+j}(\Lambda) = 0$$

for $0 \le j \le k$. From (2.3.1) it follows, in particular, that

$$V_kV_k^+\mathbf{Q}_{n+k}(\lambda) = \mathbf{Q}_{n+k}(\Lambda). \tag{7.1.20}$$

Then by induction hypothesis and by (2.1.9) we obtain from (7.1.19) that

$$\begin{aligned} V_{k+1}\mathbf{x}_{n+k+1} &= \mathbf{P}_{n+k+1}(\Lambda) \\ &+ \sum_{i=1}^{d} D_{n+k,i}^T(B_{n+k,i} + \Gamma_{1,k}A_{n+k-1,i} - V_kV_k^+B_{n+k,i}^*)V_kV_k^+\mathbf{Q}_{n+k}(\Lambda) + \ldots, \end{aligned} \tag{7.1.21}$$

where by induction it is not hard to see that the right-hand side of the above equation is a quasi-orthogonal polynomial of order $2(k+1)$. If we define $\mathbf{Q}_{n+k+1}$ to be this polynomial, then (7.1.21) becomes $V_{k+1}\mathbf{x}_{n+k+1} = \mathbf{Q}_{n+k+1}(\Lambda)$ which implies that $\mathbf{Q}_{n+k+1}(\Lambda) = 0$ and $\mathbf{x}_{n+k+1} = V_{k+1}^+\mathbf{Q}_{n+k+1}(\Lambda)$. Therefore, we only need to show that $\mathbf{Q}_{n+k+1}$ has the desired form (7.1.5); *i.e.*, the coefficient matrix of the $\mathbf{P}_{n+k}$ term is equal to $\Gamma_{1,k+1}$. We shall show that

$$\sum_{i=1}^{d} D_{n+k,i}^T(B_{n+k,i} + \Gamma_{1,k}A_{n+k-1,i} - V_kV_k^+B_{n+k,i}^*)V_k = \Gamma_{1,k+1}V_k, \tag{7.1.22}$$

which, by (7.1.20) and $\mathbf{Q}_{n+k} = \mathbf{P}_{n+k} + \ldots$, implies the desired equality. From (7.1.7) and $V_k^T(V_k^+)^T = I$ we obtain that (7.1.12) holds true, which implies, by use of $U_k^T V_k = 0$, that the second equation in (7.1.4) holds; by (2.3.1) we then conclude that also (7.1.17) holds. In particular, by $V_k^+ V_k = I$, this implies that

$$(I - V_k V_k^+)B_{n+k,i}^* = (I - V_k V_k^+)V_k V_k^T B_{n+k,i}^* = (V_k - V_k)V_k^T B_{n+k,i}^* = 0.$$

Now, by the definition of $B_{n+k,i}^*$ and by (2.1.8), the left-hand side of (7.1.22) can be written as

$$\sum_{i=1}^{d} D_{n+k,i}^T (I - V_k V_k^+) B_{n+k,i}^* V_k + \Gamma_{1,k+1} V_k = \Gamma_{1,k+1} V_k,$$

thus (7.1.22) is established. By induction, the above procedure works for all $k \leq r - 1$. We note that the last equation in (7.1.18) is

$$V_r^T A_{n+r-1,i}^T (V_{r-1}^+)^T \mathbf{Q}_{n+r-1} + V_r^+ B_{n+r,i}^* V_r \mathbf{Q}_{n+r} = \lambda_i \mathbf{Q}_{n+r},$$

which, upon multiplying by $D_{n+r,i}^T$ and doing summation for i, leads to a quasi-orthogonal polynomial of degree $n+r+1$ and order $2r+2$ which vanish on all Λ. We define this polynomial to be our $\mathbf{Q}_{n+r+1}$, and it can be shown that it is of the form of (7.1.5). Thus, we have proved that the eigenvalues of $T_{n,1}^{(r)}, \ldots, T_{n,d}^{(r)}$ are zeros of $U_k^T \mathbf{Q}_{n+k}$, $0 \leq k \leq r+1$, with $\mathbf{Q}_{n+k}$ in the form specified in the theorem. Next we prove that the matrices $T_{n,1}^{(r)}, \ldots, T_{n,d}^{(r)}$ are symmetric and mutually commuting to conclude that they have the maximal number of real eigenvalues. For the symmetry of $T_{n,i}^{(r)}$ we only need to show that for each k, $V_k^+ B_{n,k}^* V_k$ is symmetric. This follows, however, easily from (7.1.7) via $V_k^+ V_k = I$. To show that $T_{n,1}^{(r)}, \ldots, T_{n,d}^{(r)}$ are commuting under our conditions, we only need to verify (7.1.13), (7.1.14), and (7.1.15). By (2.1.7), (7.1.13) is equivalent to (7.1.6). The conditions (7.1.7) and (7.1.8) imply (7.1.4), thus, in turn, (7.1.16) and (7.1.17), from which it follows that (7.1.14) is equivalent to

$$\begin{aligned} B_{n+k,i}^* A_{n+k,j} V_{k+1} &+ A_{n+k,i} B_{n+k+1,j}^* V_{k+1} \\ &= B_{n+k,j}^* A_{n+k,i} V_{k+1} + A_{n+k,j} B_{n+k+1,i}^* V_{k+1}. \end{aligned}$$

By the definition of $B^*_{n+k,i}$ given in (7.1.1) and by (2.1.6) the above equation is equivalent to

$$\Gamma_{1,k}A_{n+k-1,i}A_{n+k,j}V_{k+1} - A_{n+k,i}A_{n+k+1,j}\Gamma_{1,k+2}$$
$$=\Gamma_{1,k}A_{n+k-1,j}A_{n+k,i}V_{k+1} - A_{n+k,j}A_{n+k+1,i}\Gamma_{1,k+2}$$

which, by (2.1.5), is true. Finally, from (7.1.16) and (7.1.17) we obtain that (7.1.15) follows from (7.1.9). Therefore, our assumptions (7.1.5)–(7.1.9) imply that $T^{(r)}_{n,1}, \ldots T^{(r)}_{n,d}$ are symmetric and commuting.

The rest of the proof follows as in Theorem 4.1.4; that the zeros are pairwise distinct and that Q is maximal will be proved in the subsequent sections. ∎

Remark 7.1.5. If we choose $U_k = I$, thus $V_k = 0$, for $k \geq 1$, then this theorem reduces to Theorem 4.1.4. However, its form is slightly different from that of Theorem 4.1.4, since we do not specify $\Gamma_{1,2}$. In general, we do not specify the $\Gamma_{j,k}$'s for $j \geq 2$. In this sense, Theorem 7.1.4 is slightly weaker. Nevertheless, we still have all $\Gamma_{1,k}$ at our disposal.

Remark 7.1.6. Again the characterization is given in terms of nonlinear equations, which are, in general, even more difficult to be solved than those in Section 4. However, because of its connection to the existence of cubature formulae and the fact that a minimal cubature formula always exists, we know that at least one solution exists; see also Remark 7.3.4 below.

Remark 7.1.7. It is likely that condition (7.1.8) is redundant. This is clear at least in one direction, for (7.1.8) is the same as the second condition in (7.1.4), which follows from the fact that Q is maximal. In the other direction, this condition may appear as a price we had to pay by switching from $S^{(r)}_{n,i}$ to $T^{(r)}_{n,i}$.

7.2. Modified Christoffel-Darboux formula

The lack of an exact formula for $A^*_{n+k-1,i}$ and for $A^{(j)}_{m,i}$ in the definition of $S^{(r)}_{n,i}$, $1 \leq i \leq d$, implies a lack of an exact formula in the recurrence relation satisfied by $\mathbf{Q}_{n+k}$, cf. (7.1.3), which makes it difficult to establish the modified Christoffel-Darboux formula for $U_0^T\mathbf{Q}_n, \ldots, U_{r+1}^T\mathbf{Q}_{n+r+1}$, in general. Fortunately, we need such a formula only for the special Q which appeared in Theorem 7.1.4. For simplicity we define

$$\text{(7.2.1)} \qquad \mathbf{K}_n^{(r)}(\mathbf{x},\mathbf{y}) = K_n(\mathbf{x},\mathbf{y}) + \sum_{k=0}^{r}[V_k^+\mathbf{Q}_{n+k}(\mathbf{x})]^T V_k^+\mathbf{Q}_{n+k}(\mathbf{y}),$$

which is consistent to (4.2.1).

Theorem 7.2.1. *Assume (7.1.6) - (7.1.9) hold, and let $\mathbf{Q}_{n+k}$ be quasi-orthogonal polynomials as in Theorem 7.1.4. Let $\mathbf{x},\mathbf{y} \in \mathbf{R}^d$. If $\mathbf{x} \neq \mathbf{y}$, then*

$$\text{(7.2.2)} \quad \mathbf{K}_n^{(r)}(\mathbf{x},\mathbf{y}) = \sum_{k=0}^{r} \frac{\mathbf{S}_{n+k,i}(\mathbf{x},\mathbf{y}) - \mathbf{S}_{n+k,i}(\mathbf{y},\mathbf{x})}{x_i - y_i}$$
$$+ \frac{(V_r^+\mathbf{Q}_{n+r}(\mathbf{y}))^T V_r^+ A_{n+r,i}\mathbf{Q}_{n+r-1}(\mathbf{x}) - (V_r^+\mathbf{Q}_{n+r}(\mathbf{x}))^T V_r^+ A_{n+r,i}\mathbf{Q}_{n+r-1}(\mathbf{y})}{x_i - y_i}$$

where for $1 \leq i \leq d$, $\mathbf{S}_{n+k,i}(\cdot,\cdot)$ is defined by

$$\begin{aligned}\text{(7.2.3)} \quad \mathbf{S}_{n+k,i}(\mathbf{x},\mathbf{y}) = (V_k^+\mathbf{Q}_{n+k}(\mathbf{y}))^T [V_k^+ A_{n+k-1,i}^{*T}(U_{k-1}^+)^T U_{k-1}^T \mathbf{Q}_{n+k-1}(\mathbf{x}) \\ + V_k^+ B_{n+k,i}^*(U_k^+)^T U_k^T \mathbf{Q}_{n+k}(\mathbf{x})] \\ + (V_{k-1}^+\mathbf{Q}_{n+k-1}(\mathbf{y}))^T V_{k-1}^+ A_{n+k-1,i}(U_k^+)^T U_k^T \mathbf{Q}_{n+k}(\mathbf{x}).\end{aligned}$$

Proof. By the proof of Theorem 7.1.4, our assumption implies that the matrices $S_{n,1}^{(r)},\ldots,S_{n,d}^{(r)}$ associated to Q are of the form of $T_{n,1}^{(r)},\ldots,T_{n,d}^{(r)}$. In particular, (7.1.10) and (7.1.11) hold. Using (7.1.10) and (2.3.1), we derive from (7.1.3) that

$$\begin{aligned}\mathbf{S}_{n+k,i}^* + V_k^T A_{n+k-1,i}^T (V_{k-1}^+)^T V_{k-1}^+ \mathbf{Q}_{n+k-1} \\ + V_k^+ B_{n+k,i}^* V_k V_k^+ \mathbf{Q}_{n+k} + V_k^+ A_{n+k,i}\mathbf{Q}_{n+k+1} = x_i V_k^+ \mathbf{Q}_{n+k},\end{aligned}$$

where

$$\mathbf{S}_{n+k,i}^* = V_k^+ A_{n+k-1,i}^{*T}(U_{k-1}^+)^T U_{k-1}^T \mathbf{Q}_{n+k-1} + V_k^+ B_{n+k,i}^*(U_k^+)^T U_k^T \mathbf{Q}_{n+k}.$$

Therefore, we obtain that

$$\begin{aligned}
\Sigma_k :=&(V_k^+\mathbf{Q}_{n+k}(\mathbf{y}))^T V_k^+ A_{n+k,i}\mathbf{Q}_{n+k+1}(\mathbf{x}) - (V_k^+\mathbf{Q}_{n+k}(\mathbf{x}))^T V_k^+ A_{n+k,i}\mathbf{Q}_{n+k+1}(\mathbf{y})\\
=&(x_i - y_i)(V_k^+\mathbf{Q}_{n+k}(\mathbf{y}))^T V_k^+\mathbf{Q}_{n+k}(\mathbf{x})\\
&-[(V_k^+\mathbf{Q}_{n+k}(\mathbf{y}))^T V_k^+ B_{n+k,i}^* V_k V_k^+\mathbf{Q}_{n+k}(\mathbf{x})\\
&\qquad - (V_k^+\mathbf{Q}_{n+k}(\mathbf{x}))^T V_k^+ B_{n+k,i}^* V_k V_k^+\mathbf{Q}_{n+k}(\mathbf{y})]\\
&-[(V_k^+\mathbf{Q}_{n+k}(\mathbf{y}))^T V_k^T A_{n+k-1,i}^T (V_{k-1}^+)^T V_{k-1}^+\mathbf{Q}_{n+k-1}(\mathbf{x})\\
&\qquad - (V_k^+\mathbf{Q}_{n+k}(\mathbf{x}))^T V_k^+ A_{n+k-1,i}^T (V_{k-1}^+)^T V_{k-1}^+\mathbf{Q}_{n+k-1}(\mathbf{y})]\\
&-[(V_k^+\mathbf{Q}_{n+k}(\mathbf{y}))^T \mathbf{S}_{n+k,i}^*(\mathbf{x}) - (V_k^+\mathbf{Q}_{n+k}(\mathbf{x}))^T \mathbf{S}_{n+k,i}^*(\mathbf{y})].
\end{aligned}$$

By (7.1.11) we have that the first square bracket in the above equation is equal to zero; and by (2.3.1) the second one can be seen to be equal to

$$\begin{aligned}
-\Sigma_{k-1}-[&(V_{k-1}^+\mathbf{Q}_{n+k-1}(\mathbf{y}))^T V_{k-1}^+ A_{n+k-1,i}(U_k^+)^T U_k^T\mathbf{Q}_{n+k}(\mathbf{x})\\
&-(V_{k-1}^+\mathbf{Q}_{n+k-1}(\mathbf{x}))^T V_{k-1}^+ A_{n+k-1,i}(U_k^+)^T U_k^T\mathbf{Q}_{n+k}(\mathbf{y})].
\end{aligned}$$

By the definitions of $\mathbf{S}_{n+k,i}^*(\cdot)$ and $\mathbf{S}_{n+k,i}(\cdot,\cdot)$, we then obtain that Σ_k can be rewritten as

$$\Sigma_k = (x_i - y_i)(V_k^+\mathbf{Q}_{n+k}(\mathbf{y}))^T V_k^+\mathbf{Q}_{n+k}(\mathbf{x}) + \Sigma_{k-1} - (\mathbf{S}_{n+k,i}(\mathbf{x},\mathbf{y}) - \mathbf{S}_{n+k,i}(\mathbf{y},\mathbf{x}))$$

In particular, since by Theorem 4.2.1 an explicit formula for Σ_0 is available, we obtain that

$$\Sigma_r = (x_i - y_i)\mathbf{K}_n^{(r)}(\mathbf{x},\mathbf{y}) - \sum_{k=0}^{r}(\mathbf{S}_{n+k,i}(\mathbf{x},\mathbf{y}) - \mathbf{S}_{n+k,i}(\mathbf{y},\mathbf{x})),$$

which, upon rearranging the terms and dividing by $x_i - y_i$ leads to the desired equation (7.2.2). ∎

If Λ is a zero of $\mathcal{Q}$, then by its definition given in (7.2.3) $\mathbf{S}_{n+k,i}(\Lambda,\mathbf{y})$ vanishes. This leads to the following corollary.

Corollary 7.2.2. *Let the assumptions be as in Theorem 7.2.1. If $\mathbf{x}_k$ and $\mathbf{x}_j$ are two zeros of $\mathcal{Q}$, then*

$$\mathbf{K}_n^{(r)}(\mathbf{x}_k,\mathbf{x}_j) = 0, \qquad \mathbf{x}_k \neq \mathbf{x}_j, \tag{7.2.4}$$

and

$$(7.2.5)\quad \mathbf{K}_n^{(r)}(\mathbf{x}_k, \mathbf{x}_k) = (V_r^+ \mathbf{Q}_{n+r}(\mathbf{x}_k))^T V_r^+ A_{n+r,i} \partial_i \mathbf{Q}_{n+r-1}(\mathbf{x}_k) + \sum_{k=0}^{r} \partial_{x_i} \mathbf{S}_{n+k,i}(\mathbf{x}_k, \mathbf{x}_k),$$

where ∂_{x_i} *means the partial derivative of* $\mathbf{S}_{n+k,i}(\mathbf{x}, \mathbf{y})$ *with respect to* x_i. *Moreover all common zeros of* $U^T \mathbf{P}_n$ and $\mathbf{Q}_{n+1}$ *are simple.*

The last statement follows from $\mathbf{K}_n^{(r)}(\mathbf{x}, \mathbf{x}) > 0$, which implies that at least one of the terms $\partial_{x_i} \mathbf{S}_{n+k,j}(\mathbf{x}_k, \mathbf{x}_k)$ or $\partial_i \mathbf{Q}_{n+r-1}(\mathbf{x}_k)$ is not zero; hence, by the definition of $\mathbf{S}_{n+k,i}$ and doing iteration we can conclude that at least one of the terms $\partial_i \mathbf{Q}_{n+k}$ is again not zero. In particular, we have

Corollary 7.2.3. *Let conditions (7.1.6) – (7.1.9) in Theorem 7.1.4 be satisfied. Then* $\mathcal{Q}$ *has* $\dim \Pi_{n-1}^d + \sigma_0 + \cdots + \sigma_r$ *many real, pairwise distinct zeros.*

7.3. Cubature formula of degree $2n-1$

We assume conditions (7.1.6) – (7.1.9), and denote the zeros of $\mathcal{Q}$ in Theorem 7.1.4 by $\mathbf{x}_1, \ldots, \mathbf{x}_N$, where $N = \dim \Pi_{n-1}^d + \sigma_0 + \cdots + \sigma_r$. Again we can consider the Lagrange interpolation based on these zeros. Using the arguments in the proof of Lemma 4.3.1, we can prove that a polynomial spanned by $V_0^+ \mathbf{Q}_n, \ldots, V_r^+ \mathbf{Q}_{n+r}$ cannot vanish on all zeros of $\mathcal{Q}$. Let

$$(7.3.1)\qquad \mathcal{V}_{n+r}^d = \Pi_{n-1}^d \cup \operatorname{span}\{V_0^+ \mathbf{Q}_n, \ldots, V_r^+ \mathbf{Q}_{n+r}\}.$$

Then we can consider the Lagrange interpolation problem by choosing a polynomial from $\mathcal{V}_{n+r}^d$ such that it interpolates an arbitrary function f at $\mathbf{x}_k$. From Corollary 7.2.2, we have that the function $L_n(f, \cdot)$, defined by

$$(7.3.2)\qquad L_n(f, \mathbf{x}) = \sum_{k=1}^{N} f(x_{k,n}) \frac{\mathbf{K}_n^{(r)}(\mathbf{x}, \mathbf{x}_k)}{\mathbf{K}_n^{(r)}(\mathbf{x}_k, \mathbf{x}_k)},$$

is a Lagrange interpolation polynomial based on the zeros of $\mathcal{Q}$. Moreover, we have

Theorem 7.3.1. *Suppose $\mathcal{Q}$ has N real, pairwise distinct zeros as in Theorem 7.1.4. Then $L_n(f)$ is the unique polynomial in $\mathcal{V}^d_{n+r}$ which interpolates f at the zeros of $\mathcal{Q}$.*

The proof of this theorem can be carried out along the lines of the proof of Theorem 4.3.2; instead of using $U^T\mathbf{P}_n(\mathbf{x}_k) = 0$ to prove (4.3.7) as in the proof of that theorem, we need to use the fact that $U_0^T\mathbb{Q}_0(\mathbf{x}_k) = 0, \ldots, U^T_{r+1}\mathbb{Q}_{n+r}(\mathbf{x}_k) = 0$, recursively, to prove that

$$\sum_{k=1}^{m} a_k \mathbf{x}_k^n = 0, \quad \ldots, \quad \sum_{k=1}^{m} a_k \mathbf{x}_k^{n+r} = 0.$$

Then using $\mathbb{Q}_{n+r+1}(\mathbf{x}_k) = 0$ and the fact that the leading terms in $\mathbb{Q}_{n+1}$ form a basis for $\Pi^d_{n+r+1}/\Pi^d_{n+r}$ finish the proof. We will not give details. We will also omit the proof of the following corollary which is in analogy to Corollary 4.3.3.

Corollary 7.3.2. *Assume (7.1.6) – (7.1.9). Then $\mathcal{Q}$ in Theorem 7.1.4 is maximal.*

Again, this corollary and Corollary 4.2.2 close the two gaps left in the proof of Theorem 7.1.4. Applying $\mathcal{L}$ to $L_n(f,\cdot)$ we obtain a cubature formula, which turns out to be of degree $2n-1$. We summarize the result in the following theorem, its proof is similar to that of Theorem 4.4.1 and will also be omitted.

Theorem 7.3.3. *Let $\mathcal{L}$ be a square positive linear functional. Suppose that $\mathcal{Q}$ in Theorem 7.1.4 has N real distinct zeros. Then $\mathcal{L}$ admits a cubature formula of degree $2n-1$ which is given explicitly as*

$$\mathcal{I}_n(f) = \sum_{k=1}^{N} \Lambda_{k,n} f(\mathbf{x}_k), \qquad \Lambda_{k,n} = [\mathbf{K}_n^{(r)}(\mathbf{x}_k,\mathbf{x}_k)]^{-1} > 0.$$

In particular, all cubature weights are positive.

Remark 7.3.4. Although the cubature formula stated in the theorem may not be minimal, every minimal cubature formula is generated by some $\mathcal{Q}$. Therefore, we can conclude that at least one set $\mathcal{Q}$ has N zeros; in particular, (7.1.6) – (7.1.9) is solvable for at least one r.

It should be pointed out that Theorem 7.1.4 and Theorem 7.3.3 are very general, but its application is mainly for small r. Indeed, since any polynomial of degree n, say P_n, can be rewritten in terms of $\{\mathbf{P}_k, 0 \le k \le n\}$ which forms a basis for Π_n^d we can consider P_n itself as a quasi-orthogonal polynomial of degree n and order n. However, if $r = n$ in Theorem 7.1.4, then we are dealing with common zeros of a general set of r_n^d many polynomials of degree n; clearly, in such a generality the equations in Theorem 7.1.4 are impossible to solve. On the other hand, for small r, the equations in Theorem 7.1.4 are tangible. One application is given in the following section, in particular, the example in Section 8.3.

8. Cubature Formulae of Even Degree

In this section we discuss cubature formulae of degree $2n-2$. In the past, formulae of even degree and odd degree have been studied separately, and it seemed that there was little similarity between them. However, as we have commented in Remark 4.1.5, the characterization of cubature formulae of even degree may appear as a special case of formulae of odd degree. This turns out to be true in general, and we shall give a characterization using our results in Section 7. We give necessary preliminaries in Section 8.1, and discuss a new characterization in Section 8.2. In particular our results indicate that from the theoretic viewpoint the construction of formulae of even degree is no easier than that of odd ones. The characterization is given in the spirit of Theorem 7.1.4 and Theorem 7.3.3. In Section 8.3 we present a nontrivial example.

8.1. Preliminaries

For a square positive linear functional $\mathcal{L}$ a cubature formula of degree $2n-2$ is a linear functional of the form (1.2.7) such that (1.2.8) holds true with Π^d_{2n-2} in place of Π^d_{2n-1}. It is known that the lower bound (1.2.9) holds in this case as well, and we again call a formula attaining the lower bound (1.2.9) Gaussian. For cubature formulae of even degree, the theory developed so far has been centered around the characterization of Gaussian formulae, which amounts to characterizing the common zeros of quasi-orthogonal polynomials $\mathbf{P}_n + \Gamma\mathbf{P}_{n-1}$. For $d = 2$ and for centrally symmetric $\mathcal{L}$ the characterization was given in [20, 27], while the proof depended heavily on algebraic ideal theory. The general case is contained in [40], in which the proof uses block Jacobi matrices as in the proof of Theorem 4.1.4. We state the following theorem using the formulation given in [40]:

Theorem 8.1.1. *The quasi-orthogonal polynomials $\mathbf{P}_n + \Gamma\mathbf{P}_{n-1}$ have $N =$*

$\dim \Pi^d_{n-1}$ *distinct zeros if, and only if, for* $1 \le i \le d$

$$(8.1.1) \qquad A_{n-1,i}\Gamma = \Gamma^T A^T_{n-1,i},$$

and, for $1 \le i, j \le d$,

$$(8.1.2) \quad \Gamma^T(A^T_{n-1,i}A_{n-1,j} - A^T_{n-1,j}A_{n-1,i})\Gamma - (A_{n-1,i}A^T_{n-1,j} - A_{n-1,j}A^T_{n-1,i})$$
$$= (B_{n-1,i}A_{n-1,j} - B_{n-1,j}A_{n-1,i})\Gamma - \Gamma^T(B_{n-1,i}A_{n-1,j} - B_{n-1,j}A_{n-1,i})^T.$$

Examples of weight functions that attain the lower bound (1.2.9), thus solving the nonlinear equations (8.1.1) and (8.1.2), are given in [3, 29] for two classes of functions, which also admit cubature formulae of degree $2n-1$. More important is the example given in [20], which states that for product Chebyshev weight of the second kind Gaussian cubature of degree $2n-2$ exists, while it is known that Gaussian cubature of degree $2n-1$ does not exist. This surprising result, however, remains singular in the sense that no other weight function, which admits cubature formulae of degree $2n-2$ but not $2n-1$, has been found. See the discussion in Section 8.3.

We shall derive in the following subsection a characterization which goes far beyond Theorem 8.1.1 and which is in the spirit of Theorem 7.1.4. To state it we need further notation.

We need to consider quasi-orthogonal polynomials of odd order. More precisely, we consider quasi-orthogonal polynomials of degree $n+k$ and of odd order; *i.e.*, polynomials in Π^d_{n+k} which are orthogonal to Π^d_{n-k-2}. *Throughout this section* we define

$$(8.1.3) \qquad \mathbf{Q}_{n+k} = \mathbf{P}_{n+k} + \Gamma_{1,k}\mathbf{P}_{n+k-1} + \ldots + \Gamma_{2k+1,k}\mathbf{P}_{n-k-1}, \quad k \ge 0.$$

For simplicity, we also write $\Gamma_0 = \Gamma_{1,0}$. The notation may seem to be in conflict with the ones given in Section 3.3 and used up to the last section; nevertheless, we emphasis here we use these new ones; moreover, when $\Gamma_{2k+1,k}$ is taken to be zero, then the notation is consistent with that of Section 3.3.

Again, we let $U_k : r^d_{n+k} \times r^d_{n+k} - \sigma_k$ be of full rank, and $V_k : r^d_{n+k} \times \sigma_k$ be matrices such that $U_k^T V_k = 0$. For $r \geq 0$ we shall study the common zeros of the polynomial set

$$(8.1.4) \qquad \mathcal{Q} = \{U_0^T \mathbb{Q}_n, \ldots, U_{r-1}^T \mathbb{Q}_{n+r-1}, \mathbb{Q}_{n+r}\};$$

for simplicity we define $U_r = I$. With these notations we adopt the definitions given in Section 3.3 in an obvious way.

8.2. Characterization

Clearly, the characterization of zeros of $\mathcal{Q}$ can be derived similarly as in Theorem 4.1.4 and Theorem 7.1.4. However, as we hinted in Remark 4.1.5, if we take $U = 0$ and $V = I$ in Theorem 4.1.4 and if we replace n by $n-1$, then Theorem 4.1.4 is exactly reduced to Theorem 8.1.1. In the same manner, the characterization of the zeros in general follows from Theorem 7.1.4. We define

$$(8.2.1) \qquad B^*_{n-1,i} = B_{n-1,i} - A_{n-1,i}\Gamma_0$$

$$(8.2.2) \qquad B^*_{n+k,i} = B_{n+k,i} + \Gamma_{1,k}A_{n+k-1,i} - A_{n+k,i}\Gamma_{1,k+1}, \quad 0 \leq k \leq r,$$

which is consistent with (7.1.1).

Theorem 8.2.1. *Let U_k, $0 \leq k \leq r-1$, be defined as above and set $U_r = I$. Then there exist quasi-orthogonal polynomials $\mathbb{Q}_{n+k}$ of order $2k$ in the form of*

$$\mathbb{Q}_{n+k} = \mathbb{P}_{n+k} + \Gamma_{1,k}\mathbb{P}_{n+k-1} + \ldots, \qquad 0 \leq k \leq r,$$

such that $\mathcal{Q} = \{U_0^T \mathbb{Q}_n, \ldots, U_r^T \mathbb{Q}_{n+r}\}$ has $\dim \Pi^d_{n-1} + \sigma_0 + \cdots + \sigma_{r-1}$ many pairwise distinct real zeros and $\mathcal{Q}$ is maximal if, and only if, Γ_k and V_k satisfy the following conditions:

$$(8.2.3) \qquad A_{n-1,i}\Gamma_0 = \Gamma_0^T A^T_{n-1,i}, \qquad 1 \leq i \leq d,$$

$$(8.2.4) \qquad B^*_{n+k,i}V_kV_k^T = V_kV_k^T B^{*T}_{n+k,i}, \qquad 0 \leq k \leq r-1, \quad 1 \leq i \leq d,$$

$$(8.2.5) \qquad U_k^T A_{n+k,i}V_{k+1} = 0, \qquad 1 \leq i \leq d, \qquad 0 \leq k \leq r-1,$$

and for $0 \le k \le r-1$ and $1 \le i,j \le d$

$$
\begin{aligned}
&V_{k-1}V_{k-1}^TA_{n+k-2,i}^T(V_{k-2}^+)^TV_{k-2}^+A_{n+k-2,j}V_{k-1}V_{k-1}^T+B_{n+k-1,i}^*V_{k-1}V_{k-1}^TB_{n+k-1,j}^{*T}\\
&\qquad +A_{n+k-1,i}V_kV_k^TA_{n+k-1,j}^T=A_{n+k-1,j}V_kV_k^TA_{n+k-1,i}^T \qquad (8.2.6)\\
&+V_{k-1}V_{k-1}^TA_{n+k-2,j}^T(V_{k-2}^+)^TV_{k-2}^+A_{n+k-2,i}V_{k-1}V_{k-1}^T+B_{n+k-1,j}^*V_{k-1}V_{k-1}^TB_{n+k-1,i}^{*T},
\end{aligned}
$$

where we take $V_r = 0$ and $V_{-1} = V_{-2} = I$.

For the proof of the theorem we only need to set $V_0 = I$ in Theorem 7.1.4 and then substitute n for $n-1$ and k for $k-1$. We remark that when the conditions (8.2.3) - (8.2.6) are satisfied, the zeros of $\mathcal{Q}$ are eigenvalues of the matrices

$$
T_{n,i}^{(r)} = \begin{bmatrix}
B_{0,i} & A_{0,i} & & & & & \bigcirc \\
A_{0,i}^T & B_{1,i} & & & & & \\
& \ddots & \ddots & \ddots & & & \\
& & B_{n-1,i}^* & A_{n-1,i}V_0 & & & \\
& & V_0^TA_{n-1,i}^T & V_0^+B_{n,i}^*V_1 & V_0^+A_{n,i}V_1 & & \\
& & & \ddots & \ddots & \ddots & \\
& & & & & V_{r-1}^+B_{n+r-1,i}^*V_{r-1} & V_{r-1}^+A_{n+r-1,i}V_r \\
\bigcirc & & & & & V_r^TA_{n+r-1,i}^T(V_{r-1}^+)^T & V_r^+B_{n+r,i}^*V_r
\end{bmatrix},
$$

for $1 \le i \le d$. When $r = 0$, Theorem 8.2.1 reduces to Theorem 8.1.1. Using the proof of Theorem 4.4.1 and noticing that common zeros of quasi-orthogonal polynomials of odd order ensures a cubature formula of degree $2n-2$ instead of $2n-1$, we can deduce the following theorem:

Theorem 8.2.2. *Let $\mathcal{L}$ be square positive. Suppose that $\mathcal{Q}$ in Theorem 8.2.1 has $N = \dim \Pi_{n-1}^d + \sigma_0 + \cdots + \sigma_{r-1}$ many real distinct zeros. Then $\mathcal{L}$ admits a cubature formula of degree $2n-2$ which is given explicitly as*

$$
\mathcal{I}_n(f) = \sum_{k=1}^{N} \Lambda_{k,n} f(\mathbf{x}_k), \qquad \Lambda_{k,n} = [\mathbf{K}_n^{(r)}(\mathbf{x}_k, \mathbf{x}_k)]^{-1} > 0.
$$

In particular, all cubature weights are positive.

The first nontrivial case beyond Theorem 8.1.1 is the case $r = 1$ in Theorem 8.2.1, which deals with the zeros of $U_0^T \mathbf{Q}_n$ and $\mathbf{Q}_{n+1}$, giving explicitly,

$$\begin{aligned} \mathbf{Q}_n &= \mathbf{P}_n + \Gamma_0 \mathbf{P}_{n-1}, \\ \mathbf{Q}_{n+1} &= \mathbf{P}_{n+1} + \Gamma_{1,1}\mathbf{P}_n + \Gamma_{1,2}\mathbf{P}_{n-1} + \Gamma_{1,3}\mathbf{P}_{n-2}. \end{aligned} \tag{8.2.7}$$

Our example in the following section will be given in this frame. We now state the theorem for this particular case in a somewhat stronger form as that of Theorem 4.1.4; *i.e.*, we give in addition the polynomial $\mathbf{Q}_{n+1}$ explicitly.

Theorem 8.2.3. *The set $\mathcal{Q} = \{U_0^T \mathbf{Q}_n, \mathbf{Q}_{n+1}\}$ has $\dim \Pi_{n-1}^d + \sigma_0$ many pairwise distinct real zeros and $\mathcal{Q}$ is maximal if, and only if,*

$$\Gamma_{1,2} = \sum_{i=1}^{d} D_{n,i}^T [(I - V_0 V_0^T) A_{n-1,i}^T + \Gamma_0 B_{n-1,i} - B_{n,i}^* \Gamma_0], \tag{8.2.8}$$

$$\Gamma_{1,3} = \sum_{i=1}^{d} D_{n,i}^T \Gamma_0 A_{n-2,i}^T, \tag{8.2.9}$$

and Γ_0, $\Gamma_{1,1}$, and V_0 satisfy the following conditions:

$$A_{n-1,i}\Gamma_0 = \Gamma_0^T A_{n-1,i}^T, \quad 1 \le i \le d, \tag{8.2.10}$$

$$B_{n,i}^* V_0 V_0^T = V_0 V_0^T B_{n,i}^{*T}, \quad 1 \le i \le d, \tag{8.2.11}$$

and for $1 \le i, j \le d$

$$\begin{aligned} &A_{n-2,i}^T A_{n-2,j} + B_{n-1,i}^* B_{n-1,j}^{*T} + A_{n-1,i} V_0 V_0^T A_{n-1,j}^T \\ &= A_{n-2,j}^T A_{n-2,i} + B_{n-1,j}^* B_{n-1,i}^{*T} + A_{n-1,j} V_0 V_0^T A_{n-1,i}^T \end{aligned} \tag{8.2.12}$$

$$\begin{aligned} &V_1 V_1^T A_{n-1,i}^T A_{n-1,j} V_1 V_1^T + B_{n,i}^* B_{n,j}^{*T} \\ &= V_1 V_1^T A_{n-1,j}^T A_{n-1,i} V_1 V_1^T + B_{n-1,j}^* B_{n-1,i}^{*T}. \end{aligned} \tag{8.2.13}$$

Proof. We first observe that in the specified case $U_1 = I$ and $V_1 = 0$, (8.2.6) is void and (8.2.10) – (8.2.13) correspond to the remaining conditions given in Theorem 8.2.1. Clearly, the proof of the formulae for $\Gamma_{1,2}$ and $\Gamma_{1,3}$ can be carried out as the one for Γ_2 in Theorem 4.1.4. Here we verify the essential part of the proof, which amounts to showing that $\Gamma_{1,2}$ and $\Gamma_{1,3}$, given in (8.2.8) and (8.2.9), satisfy the equations

$$A_{n,i}\Gamma_{1,2} = (I - V_0V_0^T)A_{n-1,i}^T + \Gamma_0 B_{n-1,i} - B_{n,i}^*\Gamma_0 \tag{8.2.14}$$

$$A_{n,i}\Gamma_{1,3} = \Gamma_0 A_{n-2,i}^T. \tag{8.2.15}$$

Indeed, since it is easy to deduce from the three-term relation (2.1.2) that

$$x_i\mathbf{Q}_n = A_{n,i}\mathbf{Q}_{n+1} + B_{n,i}^*\mathbf{Q}_n + A_{n-1,i}^{*T}\mathbf{P}_{n-1} + (\Gamma_0 A_{n-2,i}^T - A_{n,i}\Gamma_{1,3})\mathbf{P}_{n-2}$$

where

$$A_{n-1,i}^{*T} = A_{n-1,i}^T + \Gamma_0 B_{n-1,i} - A_{n,i}\Gamma_{1,2} - B_{n,i}^*\Gamma_0,$$

the equations (8.2.14) and (8.2.15) can be easily seen to imply that the zeros of $\mathcal{Q}$ are the eigenvalues of symmetric tri-diagonal matrices $T_{n,i}^{(1)}$. By Lemma 2.3.2, (8.2.15) with $\Gamma_{1,3}$ given by (8.2.9) is equivalent to

$$A_{n-1,i}\Gamma_0 A_{n-2,j}^T = A_{n-1,j}\Gamma_0 A_{n-2,i}^T, \quad 1 \le i,j \le d, \tag{8.2.16}$$

which, by the use of (2.3.2), can be seen to be equivalent to (8.2.10) (cf. [40, Lemma 4.2]). Using Lemma 2.3.2 again, (8.2.14) with $\Gamma_{1,2}$ given by (8.2.8) is equivalent to

$$\begin{aligned} &A_{n-1,i}[(I - V_0V_0^T)A_{n-1,j}^T + \Gamma_0 B_{n-1,j} - B_{n,j}^*\Gamma_0] \\ &= A_{n-1,j}[(I - V_0V_0^T)A_{n-1,i}^T + \Gamma_0 B_{n-1,i} - B_{n,i}^*\Gamma_0]. \end{aligned} \tag{8.2.17}$$

Using next the formulae of $B_{n,i}^*$ given in (8.2.2) and (2.1.6), we can verify that

$$A_{n-1,i}(\Gamma_0 B_{n-1,i} - B_{n,j}^*\Gamma_0) = B_{n-1,i}B_{n-1,j} - B_{n-1,j}^*B_{n-1,i}^* - A_{n,j}\Gamma_{1,1}\Gamma_0.$$

Substituting this equality into (8.2.17) and finally using (2.1.5) and (2.1.7), we see that (8.2.17) is equivalent to (8.2.12). Thus, (8.1.14) holds. ■

8.3. Example

From Theorem 7.1.4 and Theorem 8.2.1, it is clear that there is nothing special about cubature formulae of even degree versus odd degree from a theoretic viewpoint. However, since the matrix equations in Theorem 8.2.1 and those in Theorem 7.1.4 are somewhat different, one may be easier to solve than the other for a particular weight function, as we will now show in an example.

Again we consider the Chebyshev weight functions for $d = 2$. For the Chebyshev weight of the second kind, the equations in Theorem 8.1.1 can be solved with a simple choice

$$\Gamma = \begin{bmatrix} \bigcirc & & 1 \\ & \cdot^{\cdot^{\cdot}} & \\ 1 & & \bigcirc \\ 0 & \ldots & 0 \end{bmatrix},$$

which corresponds to the surprising result of Morrow and Patterson in [20]. For the Chebyshev weight of the first kind, solving the equations in Theorem 8.1.1 can be shown to be equivalent to solving the matrix equation

$$H^T \begin{bmatrix} 0 & -2 & & & & & \bigcirc \\ 2 & 0 & -1 & & & & \\ & 1 & 0 & & & & \\ & & \ddots & \ddots & \ddots & & \\ & & & 1 & 0 & -1 & \\ & & & & 1 & 0 & -2 \\ \bigcirc & & & & & 2 & 0 \end{bmatrix} H = \begin{bmatrix} 0 & -1 & & & & \bigcirc \\ 1 & 0 & -1 & & & \\ & \ddots & \ddots & \ddots & & \\ & & & 1 & 0 & -1 \\ \bigcirc & & & & 1 & 0 \end{bmatrix},$$

where H is a Hankel matrix (h_{i+j}). However, numerical experiments, conducted again by H. J. Schmid, indicate that this equation has no solution for $n \geq 6$. Nevertheless, it turns out that our Theorem 8.2.3 is applicable, a fact which strongly justifies the additional matrix $\Gamma_{1,1}$ in the formulae and our more general approach.

Theorem 8.3.1. *For the Chebyshev weight functions of both the first and the second kind and for even n, there exist cubature formulae of degree $2n-2$ with $\dim \Pi^2_{n-1} + n/2 + 1$ many nodes.*

Proof. We need to solve the equations in Theorem 8.2.3 for Γ_0, $\Gamma_{1,1}$ and V_0. One set of solution is given as follows:

$$\Gamma_0 = \begin{bmatrix} a^2 & 0 & a & \dots & 0 & a & 0 \\ 0 & 1 & 0 & \dots & 1 & 0 & a \\ a & 0 & 1 & \dots & 0 & 1 & 0 \\ \vdots & \vdots & \vdots & \ddots & \vdots & \vdots & \vdots \\ 0 & 1 & 0 & \dots & 1 & 0 & a \\ a & 0 & 1 & \dots & 0 & 1 & 0 \\ 0 & 1 & 0 & \dots & 1 & 0 & a \\ a^2 & 0 & a & \dots & 0 & a & 0 \end{bmatrix}, \quad \Gamma_{1,1} = \begin{bmatrix} a^2 & 0 & a & \dots & 0 & a & 0 & 0 \\ 0 & 1 & 0 & \dots & 1 & 0 & a^2 & 0 \\ a & 0 & 1 & \dots & 0 & 1 & 0 & 0 \\ \vdots & \vdots & \vdots & \ddots & \vdots & \vdots & \vdots & \vdots \\ 0 & 1 & 0 & \dots & 1 & 0 & a^2 & 0 \\ a & 0 & 1 & \dots & 0 & 1 & 0 & 0 \\ 0 & 1 & 0 & \dots & 1 & 0 & a^2 & 0 \\ a & 0 & 1 & \dots & 0 & 1 & 0 & 0 \\ 0 & a & 0 & \dots & a & 0 & a^3 & 0 \end{bmatrix},$$

where $\Gamma_0 : (n+1) \times n$ and $\Gamma_{1,1} : (n+2) \times (n+1)$, and V_0 as in (6.2.4); *i.e.*,

$$V_0^T = [a\mathbf{e}_1, \mathbf{e}_2, \dots, \mathbf{e}_m, 2\mathbf{e}_{m+1}, \mathbf{e}_m, \dots, \mathbf{e}_2, a\mathbf{e}_1], \quad n = 2m.$$

Let us verify the equations in Theorem 8.2.3 with these matrices. The condition (8.2.10) can be verified easily. Since we are dealing with centrally symmetric linear functionals, we have $B_{n,i} = 0$. In particular, we have by (2.1.7) that (8.2.12) is equivalent to

$$(8.3.1) \qquad A_{n-1,1}(\Gamma_0\Gamma_0^T + VV^T - I)A_{n-1,2} = A_{n-1,2}(\Gamma_0\Gamma_0^T + VV^T - I)A_{n-1,1},$$

which by (2.3.2) can be seen to be equivalent to $G_n^{-1}(\Gamma_0\Gamma_0^T + VV^T - I)(G_n^{-1})^T$ being a Hankel matrix. This can be easily verified, since

$$G_n^{-1}\Gamma_0^T\Gamma_0^T(G_n^{-1})^T = (a^2 + \frac{n}{2} - 1)\begin{bmatrix} 1 & 0 & 1 & \cdots & 1 & 0 & 1 \\ 0 & 1 & 0 & \cdots & 0 & 1 & 0 \\ \vdots & \vdots & \vdots & \dots & \vdots & \vdots & \vdots \\ 0 & 1 & 0 & \cdots & 0 & 1 & 0 \\ 1 & 0 & 1 & \cdots & 1 & 0 & 1 \end{bmatrix}$$

and $G_n^{-1}(G_n^{-1})^T = \operatorname{diag}\{a^{-2}, 1, \dots, 1, a^{-2}\}$. Also by $B_{n,i} = 0$ it follows that (8.2.11) then reads

$$H_i VV^T \text{ is symmetric}, \qquad H_i := \Gamma_0 A_{n-1,i} - A_{n,i}\Gamma_{1,1}, \quad i = 1, 2.$$

A careful calculation shows that $H_1 = 0$ and

$$H_2 = \begin{bmatrix} 0 & a^3 - a & \bigcirc & a - a^3 & 0 \\ -a & 0 & \bigcirc & 0 & a \\ 0 & a^2 - 1 & \bigcirc & 1 - a^2 & 0 \\ \vdots & \vdots & \vdots & \vdots & \vdots \\ 0 & a^2 - 1 & \bigcirc & 1 - a^2 & 0 \\ -a & 0 & \bigcirc & 0 & a \\ 0 & a^3 - a & \bigcirc & a - a^3 & 0 \end{bmatrix},$$

from which further calculation shows that $H_2VV^T = 0$. Using the fact that $H_1 = 0$ and $H_2VV^T = 0$, which is equivalent to $B^*_{n,1} = 0$ and $B^*_{n,2}VV^T = 0$, equation (8.2.13) reduces to equation (6.1.4) with $\Gamma_1 = 0$. Therefore, according to Example 6.2.2, we can conclude that (8.2.13) is satisfied by our choice of V and $\Gamma_{1,1}$. ∎

The cubature formulae in the theorem are generated by the common zeros of $\{U_0^T\mathbf{Q}, \mathbf{Q}_{n+1}\}$, where $\mathbf{Q}_n$ and $\mathbf{Q}_{n+1}$ are of the form (8.2.7). Since the matrix V_0 is the same as in (6.2.4) we can take the matrix U_0 as in (8.2.5). To derive the formula for $\mathbf{Q}_{n+1}$, we need, in addition to $\Gamma_{1,1}$, the matrices $\Gamma_{1,2}$ and $\Gamma_{1,3}$, which are given by the formulae (8.2.8) and (8.2.9), respectively. From the formula for $B^*_{n,i}$, which is the same as for H_i in the proof of Theorem 8.3.1, we obtain $B^*_{n,1} = 0$ and, furthermore, $B^*_{n,2}\Gamma_0 = 0$. Therefore, it is readily seen that formula (8.2.8) in this particular case reduces to formula (6.1.1). Since V_0 is the same as V in Example 6.2.2, we further conclude that $\Gamma_{1,2}$ is the same as Γ_2 in the example; that is, $\Gamma_{1,2}$ is given by the formula (6.2.7). Finally, using the formulae (6.2.6) for $D^T_{n,i}$ and (6.2.3) for $A^T_{n-2,i}$, we can use equation (8.2.9) to compute $\Gamma_{1,3}$; the result is as follows:

$$\Gamma_{1,3} = \begin{bmatrix} a^2 & 0 & a & \dots & a & 0 & a^2 \\ 0 & 1 & 0 & \dots & 0 & 1 & 0 \\ a & 0 & 1 & \dots & 1 & 0 & a \\ \vdots & \vdots & \vdots & \dots & \vdots & \vdots & \vdots \\ a & 0 & 1 & \dots & 1 & 0 & a \\ 0 & 1 & 0 & \dots & 0 & 1 & 0 \\ a & 0 & 1 & \dots & 1 & 0 & a \\ 0 & a & 0 & \dots & 0 & a & 0 \end{bmatrix}_{(n+2)\times(n-1)}.$$

With these solutions for $\Gamma_{1,1}$, $\Gamma_{1,2}$ and $\Gamma_{1,3}$, we can write down $\mathbf{Q}_{n+1}$ explicitly.

Remark 8.3.2. That equation (8.2.12) can be rewritten as (8.3.1) is true for every centrally symmetric $\mathcal{L}$ and for all dimension $d \geq 2$. Equations like this one may be of help in recognizing solutions of the matrix equations.

9. Final Discussions

In this last section we discuss two aspects of the problem. One deals with cubature formulae of degree $2n - s$, $s > 2$. The other one discusses two general aspects of constructing cubature formulae using the theorems we established in the previous sections.

9.1. Cubature formula of degree $2n - s$

For $d = 1$, Gaussian quadrature formulae have the highest degree of exactness $2n-1$; it always exists, which makes quadrature of degree $2n - 2$ less interesting from a numerical point of view. Nevertheless, because of its application in areas such as the moment problem (cf. [1, 4]) and because of its additional flexibility in placing one of the nodes, quadrature formulae of degree $2n - 2$ have received considerable attention. It seems only natural that quadrature formulae of degree $2n - s$, $s \geq 1$, have been studied as well (cf. [4, 42, 43] and the references there). In [43], for example, a characterization of $2n - s$ quadrature formulae has been given via an associated Jacobi matrix, which implies, in particular, that every positive $2n - s$ quadrature formula is Gaussian for some modified weight function.

For $d \geq 2$, the interest in cubature formulae of degree $2n-2$ is mainly due to the phenomenon characterized by the example of Morrow and Patterson; *i.e.*, there may exist Gaussian cubature formula of degree $2n-2$ but not of degree $2n-1$. However, as we discussed in the previous section, this example may be just a lucky incident. From this viewpoint, the results in Sections 7 and 8 indicate that in general there is little difference between the structure of formulae of degree $2n - 1$ and $2n - 2$; both are based on common zeros of a set of quasi-orthogonal polynomials, and both require solving nonlinear matrix equations. Looking at it from this point of view, there is little difference between formulae of degree $2n - s$, $s \geq 2$, and $2n - 1$.

By a cubature formula of degree $2n - s$, $s > 2$, we mean a linear functional in the form of (1.2.7) such that (1.2.8) holds with Π_{2n-s}^d in place of Π_{2n-1}^d. It is easy to see that such a formula is based on common zeros of

$$\mathcal{Q} = \{U_0^T \mathbb{Q}_n, \ldots, U_{r-1}^T \mathbb{Q}_{n+r-1}, \mathbb{Q}_{n+r}\}, \tag{9.1.1}$$

where the $\mathbb{Q}_{n+k}$'s are quasi-orthogonal polynomials of order $s+k-1$; for example,

$$\mathbb{Q}_n = \mathbb{P}_n + \Gamma_1 \mathbb{P}_{n-1} + \ldots + \Gamma_{s-2} \mathbb{P}_{n-s+1}. \tag{9.1.2}$$

Actually, taking $V_0 = \ldots = V_{s-2} = 0$ in the results obtained in Section 7 and substituting k by $k-s-1$ we could transform the results obtained there to the setting of $\mathcal{Q}$ defined above. In particular, we have

Theorem 9.1.1. *Let $\mathcal{L}$ be a square positive linear functional. Suppose that $\mathcal{Q}$ in (9.1.1) has $\dim \Pi_{n-1} + \sigma_0 + \cdots + \sigma_{r-1}$ distinct real zeros, which are joint eigenvalues of symmetric block Jacobi matrices. Then $\mathcal{L}$ admits a cubature formula of degree $2n-s$ with positive weights.*

We shall not formulate the theorem that characterizes the zeros of $\mathcal{Q}$ in terms of matrix equations, since the equations will be even more involved and there will be little chance to solve them in even moderate generality. To see how difficult the general situation can be, one just needs to recall that every polynomial of degree n can be viewed as a quasi-orthogonal polynomial of order $s = n$. Instead, we will give one example for which the existence of zeros can be verified directly. This example is given in terms of the symmetric polynomials which are generated by polynomials in one variable ([10, 13]); we have used a similar construction in [3, 29] to prove the existence of Gaussian cubature of both even and odd degree.

We need to recall the definition of the symmetric polynomials. A polynomial $P \in \Pi^d$ is called symmetric, if P is invariant under any permutation of its variables. In particular, the degree of P, considered as a polynomial of variable x_i, $1 \le i \le d$, remains unchanged, we denote it by $\tau(P)$. The elementary symmetric polynomials in Π^d are defined by

$$u_k := u_k(x_1, \ldots x_d) = \sum_{1 \le i_1 < \ldots < i_k \le d} x_{i_1} \cdots x_{i_k}, \qquad k = 1, 2, \ldots, d,$$

and any symmetric polynomial P can be uniquely represented as

$$\sum_{\lambda_1+\lambda_2+\ldots+\lambda_d \le \tau(P)} c_{\lambda_1,\ldots,\lambda_n} u_1^{\lambda_1} \cdots u_d^{\lambda_d};$$

where $\tau(P)$ is the degree of P.

Let $\mathbf{x} = (x_1, x_2, \ldots, x_d)$ and $\mathbf{u} = (u_1, u_2, \ldots, u_d)$, we consider the mapping $\mathbf{x} \mapsto \mathbf{u}$. The Jacobian of $\mathbf{u} = \mathbf{u}(\mathbf{x})$ can be expressed as

$$J(\mathbf{x}) := \det \frac{\partial \mathbf{u}}{\partial \mathbf{x}} = \prod_{1 \le i < j \le d} (x_i - x_j).$$

Since J^2 is a symmetric polynomial, we shall further use the notation $\Delta(\mathbf{u}) := J^2(\mathbf{x})$. Let $D = \{\mathbf{x} \in \mathbf{R}^d : x_1 < x_2 < \ldots < x_d\}$. We define R to be the image of D under the mapping $\mathbf{x} \mapsto \mathbf{u}$; *i.e.*, $R = \mathbf{u}(D)$.

Let μ be a nonnegative measure on $\mathbf{R}$ with finite moments and infinite support on $\mathbf{R}$, and let $\{p_n\}_{n=0}^{\infty}$ be the orthonormal polynomials with respect to $d\mu$. We define a measure $d\nu(\mathbf{u})$ on $R \subset \mathbf{R}^d$ by $d\nu(\mathbf{u}) = d\mu(\mathbf{x}) := d\mu(x_1) \cdots d\mu(x_d)$.

Theorem 9.1.2. *Let ν be defined as above. Then for the measures $[\Delta(\mathbf{u})]^{\frac{1}{2}} d\nu(\mathbf{u})$ and $[\Delta(\mathbf{u})]^{-\frac{1}{2}} d\nu(\mathbf{u})$ there exist cubature formulae of degree $2n - s$, $s \ge 1$, with $\dim \Pi_{n-1}^d$ many nodes. The nodes are common zeros of a $\mathbf{Q}_n$ as in (9.1.2) with $\mathbf{P}_n$ being the orthogonal polynomial with respect to the corresponding measure.*

Proof. We denote the orthonormal polynomials with respect to $[\Delta(\mathbf{u})]^{\frac{1}{2}} d\nu(\mathbf{u})$ and $[\Delta(\mathbf{u})]^{-\frac{1}{2}} d\nu(\mathbf{u})$ by $P_\alpha^{n,\frac{1}{2}}$ and $P_\alpha^{n,-\frac{1}{2}}$, respectively. For $\alpha = (\alpha_1, \ldots, \alpha_d)$, $0 \le \alpha_1 < \ldots < \alpha_d = n + d - 1$ and $n \in \mathbf{N}_0$, we have

$$P_\alpha^{n,\frac{1}{2}}(\mathbf{u}) = \frac{V_\alpha^n(\mathbf{x})}{J(\mathbf{x})},$$

where $V_\alpha^n(\mathbf{x}) = \det(p_{\alpha_i}(x_j))_{i,j=1}^d$. For $\alpha = (\alpha_1, \ldots, \alpha_d)$, $0 \le \alpha_1 \le \ldots \le \alpha_d = n$ and $n \in \mathbf{N}_0$, we have

$$P_\alpha^{n,-\frac{1}{2}}(\mathbf{u}) = \sum_{\beta \in \sigma} p_{\alpha_1}(x_{\beta_1}) \cdots p_{\alpha_d}(x_{\beta_d}),$$

where $\sum_{\beta \in \sigma}$ means summation over all permutations β of $\{1, 2, \ldots, d\}$. It is not hard to verify that $P_\alpha^{n,\pm\frac{1}{2}}$ are symmtric polynomials in $\mathbf{x}$, and are polynomials of degree n in u under the mapping $\mathbf{x} \mapsto \mathbf{u}$. For $d = 2$, these polynomials appeared in Example 2.3.

To prove the theorem, we construct quasi-orthogonal polynomials of order s, denoted by $\mathbb{Q}_n^{\pm\frac{1}{2}}$, whose zeros generate the related cubature formula. Let $q_{n,s}$ be quasi-orthogonal polynomials of order s with respect to μ; *i.e.*,

$$q_{n,s} = p_n + \rho_1 p_{n-1} + \ldots + \rho_{s-1} p_{n-s}.$$

Assume that $q_{n,s}$ has n distinct real zeros, which is always possible and in fact not unique unless $s = 1$ (cf. [4, 43]), and let the zeros be denoted and ordered by $x_{1,n} < \ldots < x_{n,n}$. For $\gamma = (\gamma_1, \ldots, \gamma_d)$, $1 \le \gamma_i \le n$ and $1 \le i \le d$, let $\mathbf{x}_{\gamma,n} = (x_{\gamma_1,n}, \ldots, x_{\gamma_d,n})$ and $\mathbf{u}_{\gamma,n} = \mathbf{u}(\mathbf{x}_{\gamma,n})$. The components of $\mathbb{Q}_n^{-\frac{1}{2}}$ are defined as follows: we define $Q_\alpha^{n,-\frac{1}{2}}(\mathbf{u})$ by replacing p_n in the definition of $P_\alpha^{n,\frac{1}{2}}$ by $q_{n,s}$ and leave p_{α_i}, $0 \le \alpha_i < n-1$, unchanged. It is not hard to verify that $Q_\alpha^{n,-\frac{1}{2}}$ is indeed a quasi-orthogonal polynomial of degree n and order s; moreover, it vanishes on $\mathbf{x}_{\gamma,n}$. Using the symmetric mapping and taking into account the restriction: $x_1 \le \ldots \le x_d$, it is readily seen that $\mathbb{Q}_n^{-\frac{1}{2}}$ vanishes on $\dim \Pi_{n-1}^d$ many distinct zeros, which implies the existence of the cubature formula of degree $2n-s$ according to Theorem 9.1.1. The components of $\mathbb{Q}_n^{\frac{1}{2}}$ are defined by

$$Q_\alpha^{n,\frac{1}{2}} = \frac{\det(q_{\alpha_i}(x_j))_{i,j=1}^d}{J(\mathbf{x})},$$

where $q_{\alpha_i} = p_{\alpha_i}$ for $1 \le i \le d-1$ and $q_{\alpha_d} = q_{n+d-1,s}$. Here we use the quasi-orthogonal polynomial in one variable $q_{n+d-1,s}$ and assume that it has $n+d-1$ many distinct zeros. The determinant $\det(q_{\alpha_i}(x_j))_{i,j=1}^d$ can be written as a sum of s determinants by splitting according to its last column, from which it follows that $\mathbb{Q}_n^{\frac{1}{2}}$ is indeed a quasi-orthogonal polynomial of degree n and order s. Moreover, it is readily seen that $Q_\alpha^{n,\frac{1}{2}}$ vanishes at $\mathbf{x}_{\gamma,n+d-1}$ and that there are $\dim \Pi_{n-1}^d$ many such points. ∎

Remark 9.1.3. For $s = 1$ and $s = 2$ this theorem means that Gaussian cubature formulae exist for the measures under consideration; the theorem has appeared in [29] for $d = 2$ and in [3] for $d \ge 2$ in these two cases. For $d = 2$, the coefficient matrices in the three-term relation (2.1.2) satisfied by orthonormal polynomials $\mathbf{P}_n^{\pm}$ are given explicitly in [29]; for $d > 2$, these matrices can be computed using various formulae given in [10].

It is interesting to notice that for any nonnegative measure μ on $\mathbf{R}$, the orthogonal polynomials corresponding to the measures $[\Delta(\mathrm{u})]^{\frac{1}{2}} d\nu(\mathrm{u})$ and $[\Delta(\mathrm{u})]^{-\frac{1}{2}} d\nu(\mathrm{u})$, respectively, possess the properties of common zeros which are analogous of the orthogonal polynomials. in several variables. These measures are the images of the product measures under the symmetric transformation $\mathbf{x} \mapsto \mathbf{u}$; both the measure and the orthogonal polynomials associated to them have simple sturctures. However, the support sets of the measures, which are the images of the product region under the symmetric transformations, are difficult to perceive.

9.2. Construction of cubature formula, afterthoughts

From the results in the previous sections, a cubature formula of degree $2n-1$ or $2n-2$ can be found by solving the matrix equations in Theorem 7.1.4 and Theorem 8.2.1, respectively. Moreover, since a minimal cubature formula exists by definition, at least one solution exists. However, the nonlinear system of equations are very difficult to solve in general, as seen by the examples. In this final section, we will discuss several aspects of constructing cubature formulae which may be of further interest.

9.2.1. First, we will comment on the connection between matrix equations in Theorem 7.1.4 and the construction of cubature formulae. We let $\mathcal{Q} = \mathcal{Q}_r$; *i.e.*,

$$\mathcal{Q}_r = \{U_0^T \mathbf{P}_n, U_1^T \mathbf{Q}_{n+1}, \ldots, U_r^T \mathbf{Q}_{n+r}, \mathbf{Q}_{n+r+1}\}.$$

If we take r as a parameter, then Theorem 7.1.4 can be viewed as a family of theorems. For a given weight function or a given linear functional $\mathcal{L}$, we can start our search for a cubature formula of degree $2n-1$ through the various parameters r. The first case is Theorem 1.1, which requires that (1.2.6) is satisfied. When (1.2.6) has no solution, we consider $\mathcal{Q}_0$. Here three groups of equations given in

(4.1.6), (4.1.7), and (4.1.8) have to be solved for V_0 and $\Gamma_{1,1}$. If it turns out that these equations do not have a solution, which means that $\mathcal{Q}_0$ does not have enough common zeros to generate a cubature formula of degree $2n-1$, then we move up to $\mathcal{Q}_1$ which corresponds to $r = 1$ in Theorem 7.1.4. In doing so, we are required to solve six groups of equations as in Theorem 7.1.4, but we also increase the degree of freedom by adding V_1 and $\Gamma_{1,2}$. Among these six groups, we have encountered two of them, namely, (7.1.6) and (7.1.7) with $k = 0$, in trying to solve the previous case of $\mathcal{Q}_0$; the third one, (7.1.9) with $k = 0$, is the third one for the case $\mathcal{Q}_0$ modified by adding one term. Moreover, as pointed out in Remark 7.1.7, there is a good chance equation (7.1.8) being redundant, one may try to put it aside first and verify it afterwards. Therefore, in terms of matrix equations, moving from $\mathcal{Q}_0$ to $\mathcal{Q}_1$ amounts to changing one of the three groups of equations and trying to solve them again in conjunction with two new groups of equations with increased degrees of freedom. In general, the extension from $\mathcal{Q}_k$ to $\mathcal{Q}_{k+1}$ works the same way. In this respect, the example in Section 8.3 explains the situation rather well. One consequence we see from this process is that in solving for $\mathcal{Q}_r$ we keep all groups of equations for $\mathcal{Q}_{r-1}$, except one, unchanged, which suggests that one should try to find as much information as possible from those equations which are left unchanged.

9.2.2. The role of the lower bound for the number of nodes of cubature formulae has been laid out in Sections 5 and 6. However, in general, we believe that the lower bound (5.2.3), or (5.4.1), will be attained only for very special weight functions; moreover, it is possible that they may be attained by different weight functions for different values of n. Even among the classical weight functions, the product Chebyshev weight of the first and the second kind are isolated in this respect. Because of the fact that lower bounds also provide lower bounds for the ranks of the V_k's in our characterization, the new improved lower bound may be of great interests. However, one may have to look for new lower bound for a restricted class of weight functions, as it is likely that a new general lower bound will be difficult to formulate, if possible at all. We explain this point through cubature of even degree in the following; the only lower bound previously known is (1.2.9).

It is not hard to see that the discussion in Section 3.2 is applicable to formulae of even degree; therefore, the lower bound (5.2.1) holds with $\sigma_0 = r_n^d - \operatorname{rank} U_0^T$ where U_0 as in Theorem 8.2.3. However, for each fixed Γ_0 we can estimate σ_0 as in the proof of Theorem 5.2.1 by considering $\mathbf{Q}_n = \mathbf{P}_n + \Gamma_0 \mathbf{P}_{n-1}$. More precisely, as an analogy to (5.2.2), we consider the subspace $\mathcal{W}$ defined by

$$\mathcal{W} = \{U_0^T \mathbf{Q}_n, \ \ x_i U_0^T \mathbf{Q}_n, \ \ 1 \le i \le d\},$$

whose rank can be estimated to give a lower bound for σ_0. Let us assume that $\mathcal{L}$ is centrally symmetric so that $B_{n,i} = 0$. In following the proof of Theorem 5.2.1, we have in place of (5.2.7) the equations

$$\sum_{i=1}^{d} \mathbf{a}_i^{*T} A_{n,i} = 0, \quad \sum_{i=1}^{d} \mathbf{a}_i^{*T} \Gamma_0 A_{n-1,i} = \mathbf{b}^T, \quad \text{and} \quad \sum_{i=1}^{d} \mathbf{a}_i^{*T} A_{n-1,i}^T = \mathbf{b}^T \Gamma_0,$$

which can be derived from the analogy of (5.2.4) by comparing the coefficients of $\mathbf{P}_{n+1}$, $\mathbf{P}_n$, and $\mathbf{P}_{n-1}$. Combining the second and the third equation gives

$$\sum_{i=1}^{d} \mathbf{a}_i^{*T} (A_{n-1,i}^T - \Gamma_0 A_{n-1,i} \Gamma_0) = 0,$$

which, together with the first one, implies that, in place of $\mathcal{A}$ in the proof of Theorem 5.2.1, we have to deal with the matrix

$$\mathcal{A} = \begin{pmatrix} A_{n,1}^T & \cdots & A_{n,d}^T \\ A_{n-1,1}^T - \Gamma_0 A_{n-1,1} \Gamma_0 & \cdots & A_{n-1,d}^T - \Gamma_0 A_{n-1,d} \Gamma_0 \end{pmatrix}.$$

We can estimate the rank of $\mathcal{A}$ just as in the proof of Theorem 5.2.1; to avoid excessive notation, we state the bound for $d = 2$ being equal to

$$\begin{aligned} \operatorname{rank} \mathcal{A} \ge & r_{n-1}^d \\ & + \operatorname{rank} \left[A_{n-1,1}(A_{n-1,2}^T - \Gamma_0 A_{n-1,2} \Gamma_0) - A_{n-1,2}(A_{n-1,1}^T - \Gamma_0 A_{n-1,1} \Gamma_0) \right]. \end{aligned}$$

Therefore, taking into consideration that $A_{n-1,i} \Gamma_0$, $i = 1, 2$, is symmetric (see (8.2.4)), we can conclude that for each fixed Γ_0 there is a lower bound for σ_0 given by

$$\sigma_0 \ge \frac{1}{2} \operatorname{rank} A(\Gamma_0),$$

where we define

$$\begin{aligned} A(\Gamma_0) = & (A_{n-1,1} A_{n-1,2}^T - A_{n-1,2} A_{n-1,1}^T) \\ & - \Gamma_0^T (A_{n-1,1}^T A_{n-1,2} - A_{n-1,2}^T - A_{n-1,1}) \Gamma_0. \end{aligned}$$

To get a general lower bound for N, we need to take the minimum among all possible Γ_0. The best we could come up with seems to be the following,

Proposition 9.2.1. *Let $d = 2$ and let $\mathcal{L}$ be centrally symmetric. Then for a cubature formula of degree $2n-2$, a lower bound for the number of nodes is*

$$(9.2.1) \qquad N \geq \dim \Pi^d_{n-1} + \frac{1}{2} \min_{\Gamma_0} \operatorname{rank} A(\Gamma_0),$$

where the minimum is taken over all Γ_0 such that (8.2.4) is satisfied.

Clearly, such a lower bound is very difficult to apply. We remark that if a Gaussian cubature formula exists, then by (8.1.2) we have $A(\Gamma_0) = 0$; therefore, the lower bound (9.2.1) coincides with (1.2.9) in this case.

References

[1] N. I. Akheizer, "The classical moment problem and some related questions in Analysis", Hafner, New York, 1965

[2] H. Berens, H. J. Schmid, Y. Xu, *On twodimensional definite orthogonal systems and on a lower bound for the number of nodes of associated cubature formulae*, SIAM J. Math. Anal. (to appear).

[3] H. Berens, H. J. Schmid, Y. Xu, *Multivariate Gaussian cubature formulae*, Arch. Math. (to appear).

[4] T. S. Chihara, "An introduction to orthogonal polynomials", Mathematics and its Applications, Vol. 13, Gordon and Breach, New York, 1978.

[5] R. Cools and H. J. Schmid, *Minimal cubature formulae of degree $2k-1$ for two classical functions*, Computing, **43** (1989), 141-157.

[6] P. J. Davis and P. Rabinowitz: "Methods of Numerical Integration", Academic Press, New York, 1975.

[7] A. Erdélyi, et al, "Higher Transcendental Functions", Vol. 2, McGraw-Hill, New York, 1953.

[8] H. Engles, "Numerical Quadrature and Cubature", Academic Press, New York, 1980.

[9] R. A. Horn and C. R. Johnson, "Matrix Analysis" Cambridge University Press, 1985.

[10] S. Karlin and J. McGregor, *Some properties of determinates of orthogonal polynomials*, in "Theory and Applications of Special Functions" , R. A. Askey (ed.), Academic Press, 1975.

[11] M. A. Kowalski, *The recursion formulas for orthogonal polynomials in n variables*, SIAM J. Math. Anal. **13** (1982), 309-315.

[12] M. A. Kowalski, *Orthogonality and recursion formulas for polynomials in n variables*, SIAM J. Math. Anal. **13** (1982), 316-323.

[13] T. Koornwinder, *Two-variable analogues of the classical orthogonal polynomials*, in "Theory and Applications of Special Functions", R. A. Askey ed., Academic Press, 1975.

[14] H. L. Krall and I. M. Sheffer, *Orthogonal polynomials in two variables*, Ann. Mat. Pura. Appl. (4) **76** (1967), 325-376.

[15] V. A. Kuz'menkov, *The existence of cubature formulas with the least possible number of node* Zh. vychisl. Mat. mat. Fiz. **16** (1976), 1337 - 1339.

[16] R. A. Lorentz, "Multivariate Birkhoff Interpolation", Lecture Notes in Mathematics, No. 1516, Springer Verlag, 1992.

[17] H. M. Möller, *Polynomideale und Kubaturformeln*, Thesis, Univ. Dortmund, 1973.

[18] H. M. Möller, *Kubaturformeln mit minimaler Knotenzahl*, Numer. Math. **25** (1976), 185-200.

[19] H. M. Möller, *Lower bounds for the number of nodes in cubature formulae*, in "Numerical Integration", ISNM Vol. 45, G. Hämmerlin (ed), Birkhäuser, Basel, 1979.

[20] C. R. Morrow and T. N. L. Patterson, *Construction of algebraic cubature rules using polynomial ideal theory*, SIAM J. Numer. Anal. **15** (1978), 953-976.

[21] I. P. Mysovskikh, *Numerical characteristics of orthogonal polynomials in two variables*, Vestnik Leningrad Univ. Math. **3** (1976), 323-332.

[22] I. P. Mysovskikh, *The approximation of multiple integrals by using interpolatory cubature formulae*, in "Quantitative Approximation", R. A. DeVore and K. Scherer (eds.), Academic Press, New York, 1980, 217-243.

[23] I. P. Mysovskikh, "Interpolatory cubature formulas", "Nauka", Moscow, 1981.

[24] I. P. Mysovskikh and V. Ya. Chernitsina, *Answer to a question of Radon*, Dokl. Akad. Nauk SSSR **198** (1971), **3**, 537-539.

[25] G. Renner, Darstellung von strickt quadratpositiven linearen Functionalen auf endlich-dimensionalen Polynomraumen, Dissertation, University of Erlangen-Nuremberg, 1986, pp 93.

[26] J. Radon, *Zur mechanischen Kubatur*, Monatsh. Math. **52** (1948), 286-300.

[27] H. J. Schmid, On cubature formula with a minimal number of knots, Numer.

Math. **31** (1978), 282-297.

[28] H. J. Schmid, *Interpolatorische Kubaturformeln*, Diss. Math. **CCXX**, 1983, 1-122.

[29] H. J. Schmid and Y. Xu, *On Bivariate Gaussian cubature formula*, Proc. Amer. Math. Soc. (1994), in print.

[30] A. Stroud, "Approximate Calculation of Multiple Integrals", Prentice Hall, Englewood Cliffs, New Jersey, 1971.

[31] P. K. Suetin, "Orthogonal Polynomials in Two Variables", "Nauka", Moscow, 1988.

[32] G. Szegö, "Orthogonal Polynomials," Amer. Math. Soc. Colloq. Publ. Vol.23, Providence, 4th edition, 1975.

[33] P. Verlinden and R. Cools, *On cubature formulae of degree $4k+1$ attaining Möller's lower bound for integrals with circular symmetry*, Numer. Math, **61** (1992), 395-407.

[34] Y. Xu, *On multivariate orthogonal polynomials*, SIAM J. Math. Anal. **24** (1993), 783-794.

[35] Y. Xu, *Gaussian cubature and bivariate polynomial interpolation*, Math. Comp. **59** (1992), 547-555.

[36] Y. Xu, *Unbounded commuting operators and multivariate orthogonal polynomials*, Proc. Amer. Math. Soc. **119** (1993), 1223-1231.

[37] Y. Xu, *Multivariate orthogonal polynomials and operator theory*, Trans. Amer. Math. Soc. **343** (1994), 193-202.

[38] Y. Xu, *Block Jacobi matrices and zeros of multivariate orthogonal polynomials*, Trans. Amer. Math. Soc. **342** (1994), 855-866.

[39] Y. Xu, *Recurrence formulas for multivariate orthogonal polynomials*, Math. Comp. **62** (1994), 687-702.

[40] Y. Xu, *On zeros of multivariate quasi-orthogonal polynomials and Gaussian cubature formulae*, SIAM J. Math. Anal. **25** (1994), 991-1001.

[41] Y. Xu, *Solutions of three-term relations in several variables*, Proc. Amer. Math. Soc. (1994), in print.

[42] Y. Xu, *Quasi-orthogonal polynomials, quadrature, and interpolation*, J. Math.

Anal. Appl. **182** (1994), 779-799.

[43] Y. Xu, *A characterization of positive quadrature formulae*, Math. Comp. **62** (1994), 703-718.

[44] Y. Xu, *On a class of bivariate orthogonal polynomials and cubature formula*, Numer. Math. (to appear).

Anal. Appl. 182 (1994), 79-89.

(3) Y. Xu, A characterization of positive quadrature formulae, Math. Comp. 62 (1994), 703-718.

(4) Y. Xu, On a class of bivariate orthogonal polynomials and cubature formula, Numer. Math. (to appear).